# Introduction

This Panzer Tracts No. 6-3 covers the history of the development, evolution, production, modification, and testing of the **Panzerkampfwagen Maus** and **E 100** (along with its predecessor the **Tiger-Maus**). As is our high standard, Panzer Tracts are based solely on surviving specimens, wartime photographs, and the content of primary source documents written by those who participated in the design, production, and employment of the Panzers. The time is long overdue for the real experts who designed and produced these super-heavy Panzers to have their say.

Knowing the role of the players involved is key to understanding the development history. As used in this history, the names Krupp and Porsche (without titles) refer to the firms or their representatives and not the founding fathers.

Before 1932, the German Army had established a well-controlled weapons development and procurement system involving Inspecktorat 6 (In 6), Waffenpruefen 6 (Wa Pruef 6), and contractors. In 6 was responsible for creating performance requirements for new vehicles wanted by the troops. In 6 also approved the final designs for series production. Wa Pruef 6 created the design specifications, awarded design contracts, and held meetings to control the projects. The commercial designers were informed of the specifications that they had to meet and which components to use. Firms like Daimler-Benz, Krupp, and Rheinmetall were awarded contracts by Wa Pruef 6 to develop the detailed designs and produce test vehicles. The results were then inspected by In 6 prior to approving further development.

This well-regulated procurement system with its built-in checks and balances worked well in creating the **Pz.Kpfw.I**, **II**, **III**, and **IV**, successfully employed in campaigns of conquest from 1939 to 1942. However, the wheels had already started falling off in 1938 when Oberbaurat Kniepkamp in Wa Pruef 6 was allowed to develop and produce high-speed Panzers (when there was no tactical requirement for high speed at the cost of reliability). Then the politicians became involved in tank design with the creation of the Panzerkommission headed by Prof. Dr. Porsche in 1939. The end result was a design, development, and procurement system that was out of control with politicians demanding that the German army have heavy tanks that were superior to the enemy's.

The resulting heavy Panzer designs were no longer based on sound functional requirements that were interpreted and strictly enforced by design specifications. Instead, anyone at any given meeting could add novel ideas as requirements. The designers would then quickly throw together a makeshift design for presentation at the next meeting where it could be altered, approved, or abolished at the whim of any bureaucrat or officer.

Examples of the unique ideas that were seriously considered and designed for the Maus project include: (1) twin flamethrowers mounted at the rear remotely controlled by the radio operator as directed by the commander, (2) a vertical 2 cm anti-aircraft gun rigidly mounted in the turret, which was to be fired at low-flying strafing and bombing aircraft as they passed overhead, (3) a submersion kit designed for the **Maus** to ford 6 meter deep bodies of water in which the 50 mm thick engine compartment deck plates with air intake louvers had to be replaced - but no lifting method was provided, and (4) 180 mm thick side armor plates in which 80 mm had to be milled out (a total of 4.5 metric tons wasted per plate) to create a lower section that was 100 mm thick.

This out of control situation was recorded in detail in Krupp's Maus project meeting minutes and correspondence files, which survived thanks to the School of Tank Technology hauling them off to England at the end of the war. Excerpts from these primary source documents have been used to create this history of the development of the **Maus** and **Tiger-Maus (E 100)**.

Thanks to a private collector, the detailed large-scale component drawings created by Porsche and Krupp and used by Alkett for Maus assembly have survived. In addition, large scale turret drawings from Krupp as well as a large scale drawing from Adler of the **E 100** were found in archives. These original drawings were used to create the new scale drawings of the **Maus** and **E 100** contained in this Panzer Tracts.

As we have found in numerous cases, the chassis designers did not keep up with turret developments and used outdated turret designs on their overview drawings. This is also true for both Porsche and Adler. Adler drew an out of date **Maus** turret from 1942 on the **E 100** drawing which they completed for the Allies from partially burnt drawings directly after the war. It was pure luck that evidence was found that a turret was designed by Krupp for the **E 100** in 1944 which, with the exception of the armor thicknesses, was the same as the **Maus II** turret. And, even greater luck that an original drawing of a **Maus II** turret was found. Almost forty years of digging in archives has paid off with this and other rare finds.

P.S. It is a common belief that the **Maus** weighed too much and therefore the project was cancelled because of the lack of resources. This is pure bunk. Contracts were placed by Wa Pruef 6 for six **Pz.Kpfw.Maus** as a trial series, followed by contracts from Wa J Rue (WuG 6) for 135 **Pz.Kpfw.Maus** as a mass production series. Production was well underway with armor plates already rolled for 30 **Maus** hulls and cut for nine hulls and turrets when disaster struck at Krupp, Essen in early August 1943. This is the only incident in the war where a bombing raid succeeded in completely stopping mass production of a Panzer.

## CONCEPTUAL DESIGN EVOLUTION

The earliest notes revealing information on plans to design a super-heavy Panzer are found in the minutes of Hitler's conferences with Albert Speer, Reichsminister fuer Bewaffnung und Munition (head of the ministry for armaments and ammunition) as follows:

<u>5-6 March 1942, Item 2</u> - *Directive to Krupp that instead of a **72 ton Panzer**, a **100 ton Panzer** is to be rapidly developed as a trial vehicle. The first trial vehicle should be operational in the shortest time, in all cases before the Spring of 1943.*

<u>21-22 March 1942, Item 18</u> - *Porsche is to be given the contract for independent design of a **100 ton Panzer**.*

<u>14-15 April 1942, Item 10</u> - *At least 100 rounds of ammunition are to be carried in the **100 ton Panzer**. It should have a machinegun in addition to the heavy gun but not a lighter quick-firing gun. Remote control of the machinegun is quite acceptable to avoid a hole in the armor plate.*

The turret specifications for this **100 ton Panzer** were discussed during an internal meeting at Krupp on 18 April 1942 as follows: *A new proposal is to be created for a turret with a **15 cm L/40** gun ("L/40" is the caliber length of the gun -- 15 cm x 40 equals 6.00 meters) with cartridges instead of separate two-piece ammunition in order to achieve a rate of fire of 4 to 5 rounds per minute. In addition, the projectile weight is to be reduced from 43 to 34 kilograms with an associated increase in the muzzle velocity to about 845 meters per second. Part of the ammunition is to be stowed in a backpack on the turret, out of which one should be able to load the gun at elevations from -8 to +15°. In addition, an attempt should be made to achieve an elevation of up to 40° through 360° traverse. The assumption is approved that the Panzer be driven into a position where the gun can be loaded out of the backpack. This turret is to be offered to the Porsche firm for their **VK 100.01** by 15 May. In addition, we should determine if it is more favorable to build a turret with a **12.8 cm L/50** gun firing a 29.3 kilogram projectile with a muzzle velocity of 810 meters per second.*

As discussed in Hitler's conferences with Speer:

<u>13 May 1942, Item 28</u> - *Hitler emphasized that it must be calculated that the heaviest Russian tanks will certainly appear by Spring. Therefore he demands that the heavy Panzers currently being designed be energetically carried out and holds the opinion that reducing the weight to 70 tons is incorrect. He has no qualms that instead of 100 tons one could even get up to a weight of 120 tons. Priority is to be given to the heaviest armor connected with a gun with the highest performance. From the start, he wants a gun with a length of L/60 or eventually even L/72.*

<u>4 June 1942, Item 40</u> - *Hitler is in agreement that the super-heavy Panzer be a slow-moving vehicle (mobile fortress).*

<u>23 June 1942, Items 37 and 38</u> - *Hitler has approved the drawings of the heavy Porsche Panzer with several modifications, including strengthening the belly to 100 mm and alternatively a **15 cm L/37** or a **10.5 cm L/70** gun. Hitler favors the 10.5 cm gun because of the higher rate of fire, greater ammunition stowage, and better ability to serve the gun. However, he believes that plans can be made for both guns to be selectively mounted in this type of large Panzer. He doesn't consider it necessary to have a secondary turret with a 7.5 cm gun, because escorting Panzers must be assigned. He is satisfied with the proposal and in agreement with the model. Professor Porsche promised delivery of the first vehicle by 12 May 1943.*

*Hitler agrees with the principles for designing a Panzer that first priority is the heaviest armament, second priority high speed, and third priority heavy armor. However, he also believes that heavy armor is unavoidably necessary.*

In response to Krupp's proposal dated 25 June, OKH/Wa Pruef 6 awarded contract SS 006-4467/42 to Fried.Krupp A.G., Abt.A.K., Essen for designing a turret for the **Pz.Kpfw. "Maeuschen"** on 17 July 1942. At this time, the design specifications were for a turret with 360 degree traverse constructed out of rolled armor plate (250 mm front, 200 sides and rear, 80 mm roof) with a cast gun mantle. Two guns were to be mounted with an elevation arc from -7 to +25 degrees, a **15 cm Kw.K. L/31** with a maximum range of 16 km firing a 43.5 kg round at a muzzle velocity of 750 m/sec penetrating 190 mm at 30° at 1000 meters range, and a **7.5 cm Kw.K. L/24** with a maximum range of 7 kilometers. Elevation was by hand and traverse by electric drive and hand. A total of 25 rounds or 15 cm ammunition and 50 rounds of 7.5 cm ammunition were to be stowed in the turret weighing 57 tons.

Conceptual design drawings were prepared, and details associated with designing a Panzer that could fit inside the rail transport profile were discussed between Krupp and Porsche on 27 August 1942:

*If the Panzer is loaded on the railcar so high that the outer tracks don't exceed the specified profile, there isn't sufficient space for the turret. The drawing of the turret shown on the loading diagram isn't correct because the 3100 mm width as shown can only be achieved by milling off the turret side walls to a height of 200 mm. Also, the commander's cupola can't be relocated to the middle because it won't be usable from inside. Krupp proposes a redesign so that the Panzer is transported between two rail cars with its tracks barely clearing the rail line.*

*Other changes to Porsche drawing 1434 included: The gun mantle has been designed so that cutouts on the hull deck are not needed. In addition, the turret can be raised about 50 mm to allow a cylindrical socket for a water sealing band. On a suitable position on the turret, a flange can be located for mounting a **Luftschacht fuer Unterwasserfahrt** (air chamber for submersion) with 800* mm internal diameter. *The **Turmtragring** (turret race) extending into the side walls is superseded by the above transport proposal. Therefore the side walls of the hull can extend vertically up to the height of the deck.*

On 22 September 1942, OKH/Wa Pruef 4 informed Krupp: A*s a result of a new decision, the turret for the "Loewe" (10.5 cm L/70) (refer to Panzer Tracts 20-1) is no longer to be produced. Instead it is to be replaced by a turret for the "Maus" with **15 cm L/37** and **7.5 cm L/24** guns for which Krupp is hereby awarded contract SS-004/8015/42 for 1500 designer man-days.*

Porsche prepared conceptual design drawing Sk.7949 dated 5 October 1942 (amended 12Oct42) for their **Typ 205 A** listing both a **15 cm L/37** and a 12.8 cm gun. While normal crew access was through the turret, there were three emergency hatches: one in the turret rear and two at the front on the hull deck. A total of 22 rounds of 12.8 cm and 45 rounds of 7.5 cm ammunition were stowed in the turret and an additional 20 rounds of 12.8 cm and 50 rounds of 7.5 cm in the hull. One **M.G.34** was mounted in the hull front. The vehicle's total weight of 150 metric tons was to be carried on two 500 mm wide (or one 1000 mm) tracks with a pitch of 130 mm. Track contact length of 4.60 meters (increased to 4.9 m when sunk in) resulted in a very high ground pressure of 1.54 kg/cm2. The suspension consisted of 12 double roadwheels. The drive train was electrical, designed to provide a maximum speed of 20 km/hr and a minimum speed of 1.5 km/hr. A water-cooled 44.5 liter 12-cylinder Daimler-Benz diesel engine (rated at 1000 metric horsepower at 2400 rpm) drove a generator that provided power to two electric motors designed to provide a maximum speed of 20 km/hr and a minimum speed of 1.5 km/hr. With a rear drive sprocket (918 mm dia.), it could climb slopes up to 25° (47%).

On drawing SK.7948 dated 5 October 1942, Porsche proposed a second model, **Typ 205 B**. With the exception of the engine it was exactly the same as the **Typ 205 A**. **Typ 205 B** was to have an air-cooled 41.5 liter **Typ 205/2** Porsche diesel rated at 780 metric horsepower at 2000 rpm.

On 5 November 1942, Porsche sent drawing E 1549 and 205.00.201 to Krupp for comment. These conceptual drawings incorporated Krupp's turret drawing Bz 2519 dated 4Sep42. *Porsche had reduced the armor thickness of the hull on average 10% in order to reduce the vehicle weight. To meet the total vehicle weight of 168 metric tons, Porsche asked Krupp to reduce armor thickness of the turret in order to decrease its total weight with ammunition from 47 tons to 43 tons. Porsche was forced to take these measures because of the relatively high ground pressure of the tracks. Lowering the ground pressure by extending the track width wasn't possible because of the rail loading profile and the interior space couldn't be re-*

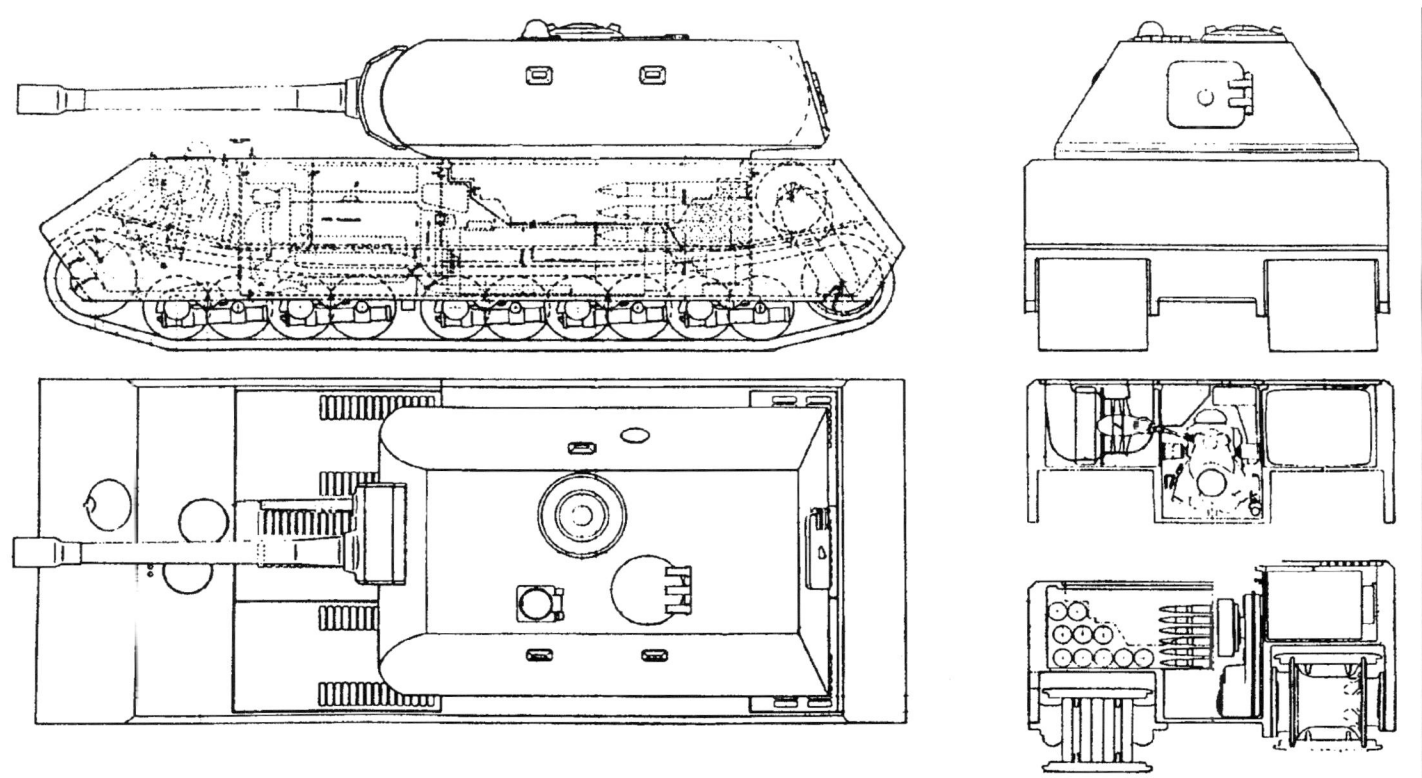

**Above: An original Porsche drawing from 1Jan43 of a conceptual design for a Maus, which has many features that were dropped during the design evolution in early 1943, including the hull machinegun, the round hatches for the driver and radio operator, the commander's cupola, vision ports on the turret sides, an escape hatch on the turret rear, and the Porsche torsion bar suspension. (TTM)**

duced. It also couldn't be reduced by extending the track contact length, which had been stretched to the extreme - governed by the steering ability.

Conceptual designs for the **"Maeuschen"** were discussed at a meeting at the Min.f.Bew.u.Mun. (Odl. Saur) attended by the Panzerkommission (General v. Radelmeier, Prof.Dr.Porsche), In 6 (Oberst Thomale, Oberstlt. Bollbrinker), Pruef 6 (Oberst v. Wilcke, Oberstlt. Crohn, Oberbaurat Kniepkamp) on 17 November 1942:

*As already requested at a meeting of the Panzerkommission on 10 November, Krupp was to prepare an alternative proposal with a turret mounted in the rear. The Krupp proposal is to be quickly completed. A decision is to be made in three to four weeks whether the Porsche or Krupp proposal is to be produced.* (Refer to the text section on Krupp's **Tiger Maus/E 100**).

*Porsche presented their proposal for a* **170 ton Panzer** *with the turret in the rear. The engine power of 900 horsepower can be increased to 1500 by a compressor mounted on the water-cooled Daimler-Benz gasoline Motor 603. Aircraft engines are not robust enough for Panzers. Porsche has installed the engine in front of the turret, the electric generator under it, and two electric motors at the rear. The electrical machines are taken over from the* **Tiger (P)** (refer to **Panzerkampfwagen VI P** from Panzer Tracts) *but with higher revolutions. The 8 roadwheels per side have longitudinal torsion bar suspension. Track contact length is 5.72 meters with an overall vehicle length of 8.85 meters. Firing height is 2.7 meters. The armor thicknesses are 200 mm front, 180 mm upper sides, 100 mm lower sides, 180 mm rear, 100 mm belly front, 50 mm belly rear.*

*A* **Sturmgeschuetz** *design was suggested to accelerate production. However, this was turned down by Oberst Thomale because a* **Sturmgeschuetz** *was unsuitable based on tactical considerations. The* **Maeuschen** *is to be assigned the mission assisting the infantry by slugging through the enemy defensive positions. In addition, a large saving in development time is not expected.* (Refer to the following text section on the **Maus/E 100 Sturmgeschuetz**).

As discussed in Hitler's conferences with Speer on 1-3 December 1942, Items 14 and 15: *Hitler took great interest in the presentations of Professor Porsche and Dr. Mueller (Krupp) on the preliminary work on the* **"Maeuschen"**. *He expects completion of the first trial vehicle in the Summer of 1943 followed by production of five each month. These vehicles are to be assembled by Krupp.*

*For the super heavy Panzer (***Maeuschen***), Hitler wants information of the penetrating ability of the 15 cm gun, the 12.7 cm naval gun, the 12.8 cm Flak gun, and a 12.8 cm gun with the greatest caliber length. Hitler also requested a review of the Navy's inventory of armor plates by thickness that can be used for the super-heavy Panzer.*

On 10 December 1942, Oberstlt. Mueller from Wa Pruef 6 and Obering. Dorn and Dr. Bankwitz from Krupp met to discuss the **Maus Turm** on drawing Bz 2599 that was intended for both the **Porsche Maus** and **Krupp Tiger-Maus**, resulting in the following notes:

*1. Wa Pruef 6 will decide if the armor thickness can be reduced from 250 mm front and 200 sides to 225 mm front and 180 mm sides in order to lower the total weight.*
*2. Krupp is to immediately begin on the 1:1 scale wooden model. Wa Pruef 6 agrees with the orientation of the weapons in the turret. For elevation the machinegun should be coupled to the gun carriage.*
*3. Wa Pruef 6 agrees with the* **Zielfernrohr** *(periscopic gunsight) extending through the roof. Attempt to move it somewhat further forward.*
*4. Sealing the gunsight and other smaller penetrations with outer caps is allowed.*
*5. The* **Spiegelkuppel** *(cupola with periscopes) planned for observation is approved.*
*6. Wa Pruef 6 agrees with the crew hatch on the turret roof and the chassis deck being the same; the hatch on the* **Kommandantenkuppel** *can be a different design. Wa Pruef 6 is in agreement with the location of the crew members. The adjustable seat designed by Wegmann should be tried for the commander. If there isn't space, plan to install a height-adjustable motorcycle saddle for both standing and sitting positions.*
*7. Wa Pruef 6 agrees with the location and shape of the elevation and traverse gear as well as the auxiliary traverse.*
*8. Plan space for a compressed air tank to blow fumes out of the spent cartridges after firing.*
*9. Krupp added that the same turret is planned for both the Krupp and Porsche* **Maus**.
*10. Wa Pruef 6 agrees with the positioning of a simple hydraulic jack to raise the turret about 5 mm for firing and to lower the turret onto a seal.*
*11. Determine if an* **Entfernungsmesser** *(range finder) can be installed in the front of the turret. A* **Nebelwurfgeraet** *(smoke grenade discharger) is planned for on the turret roof. Since these can also fire explosive grenades, Krupp recommends that one or two additional* **Nebelwurfgeraet** *be installed toward the rear of the turret roof if space is available.*
*12. Since a penetration for a vision slit in such thick armor is difficult, Krupp recommended a Busch or Zeiss* **Winkelfernrohr** *(periscopic sight).*

On 15 December 1942, Krupp was informed by Wa Pruef that work on their **Maus** project (a weaker model with **Tiger** components) was to be halted because, following discussions with Chef H Rust, Oberstlt. Holzheuer has decided that only a **Porsche-Maus** with a **Krupp-Turm** was to be produced.

On 21 December 1942, the armor thicknesses on

the **Typ 205 "Maus" Wanne** (hull) were listed as: 200 mm front, 80 mm inner sides, 100 mm outer sides, 150 mm rear, 100 mm front deck and 50 mm rear deck, 100 mm front belly, and 50 mm rear belly.

Hull design details were discussed at a meeting at Porsche in Zufferhausen with Wa Pruef and Krupp on 23 December 1942: *The hull deck is to be raised so that the so-called **Turmkragen** (turret collar) can be dropped. It is difficult to fit the current design of the **M.G.42** into a **Kugelblende** (ball mount). Either the machinegun must be pulled back so that only the muzzle sticks out or a new ball must be designed if the machinegun must stick farther out. The tunnel between the turret and the driver's compartment will be difficult to design. 100 rounds of 7.5 cm ammunition are to be stowed in the hull so that only 25 to 30 rounds are stowed in the turret.*

The decision to produce the **Porsche-Maus** was made by Hitler in a conference with Speer on 3-5 January 1943: *After thorough consideration and comparison of all the advantages and disadvantages of the Krupp and Porsche proposals for the **"Maeuschen"**, Hitler decided that the Porsche proposal be accepted. Porsche is responsible for the design of the vehicle, Krupp for the production of the hulls and turrets, and Alkett for the assembly. Production of 10 per month is the final goal. Completion of the first vehicle and start of production must be strived for the end of 1943.*

*Based on the report on the situation of the armor-piercing ammunition, Hitler maintains that the 12.8 cm gun is the most suitable gun for the **"Maeuschen"**. However, in addition, a turret with a 15 cm gun is to be projected. Firing trials with the 12.8 cm gun are to be immediately conducted with shaped charge, tungsten core, 12.8/8.8 cm and 12.8/10.5 cm discarding sabot and eventually also with different types of propellants. Because the 12.8 cm Flak gun with its sectional design can't be used without modification, tests are to be conducted to determine if a 12.8 cm gun with a caliber length of L/70 is usable instead of L/61.*

*Technical superiority can only be assured for a combat period of one year at most. Therefore one must now already plan for achieving superiority in 1944. For this year the **Tiger** and **Panther** are superior. The **"Maeuschen"** and the new **Tiger** with the **8.8 cm L/71** gun must bring this superiority for 1944.*

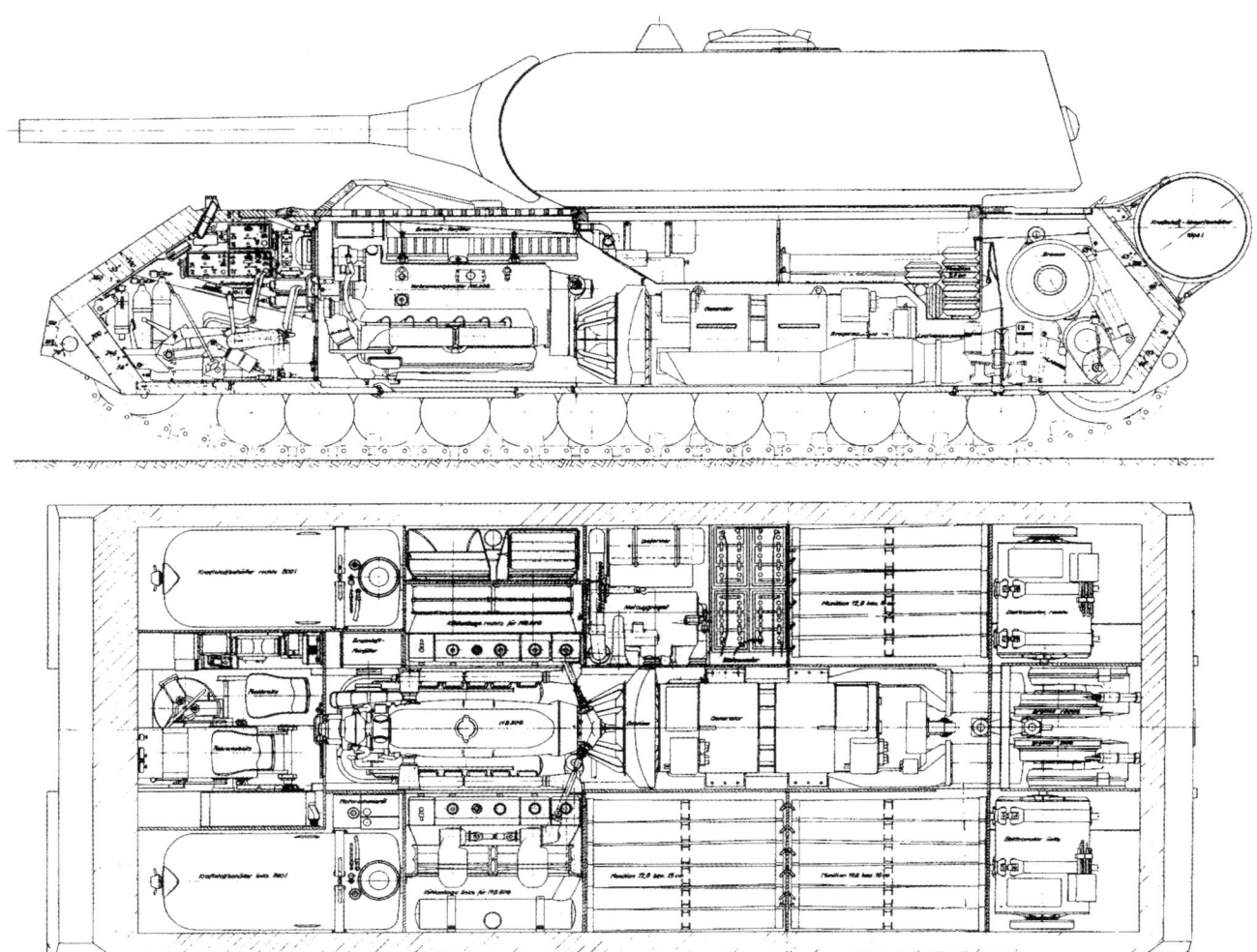

**Above: An original Porsche drawing of the Maus chassis interior layout with stowage for single-piece 12.8 cm rounds. Porsche copied an outdated turret drawing from Krupp that still has a cupola. (TTM)**

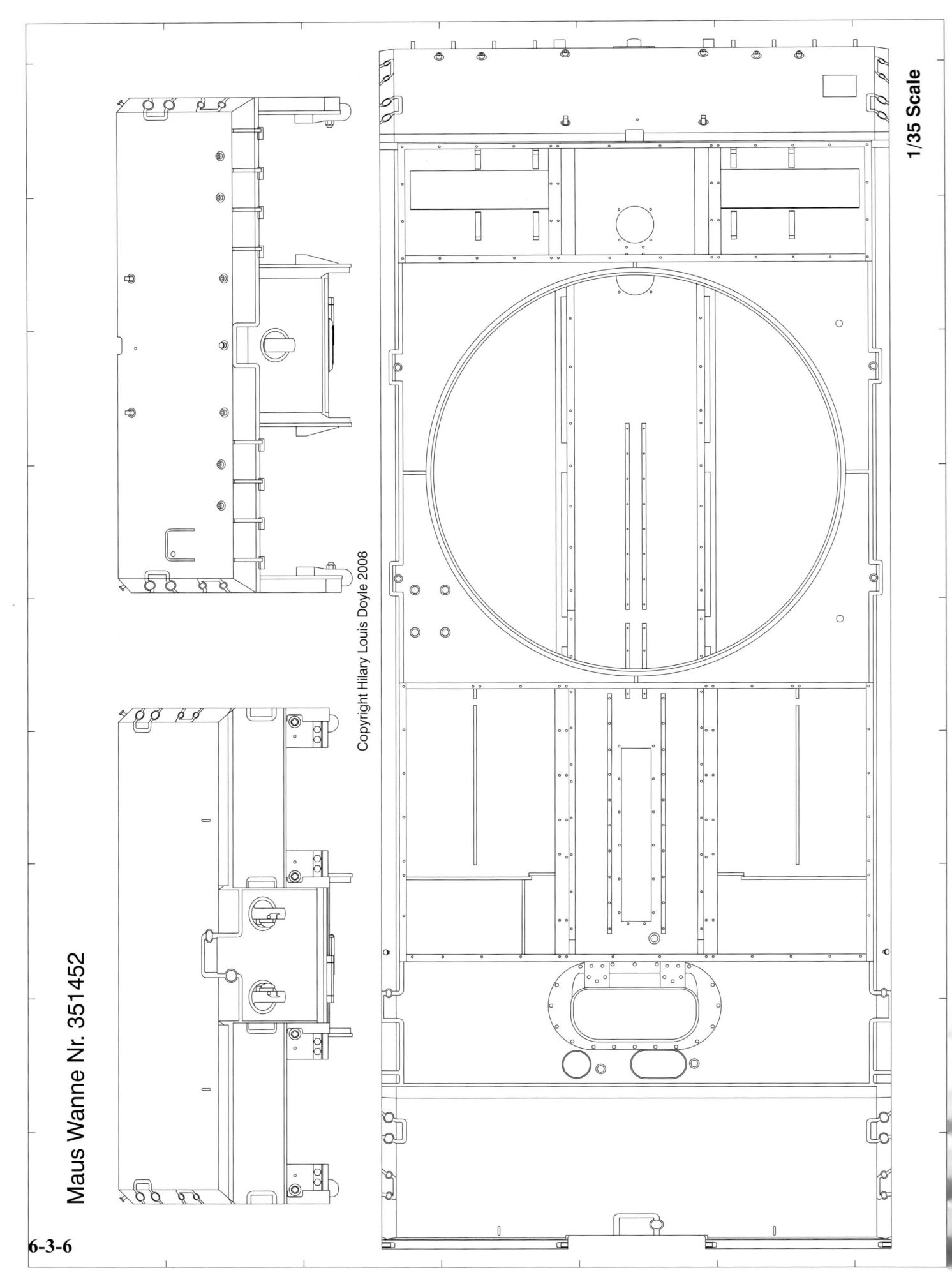

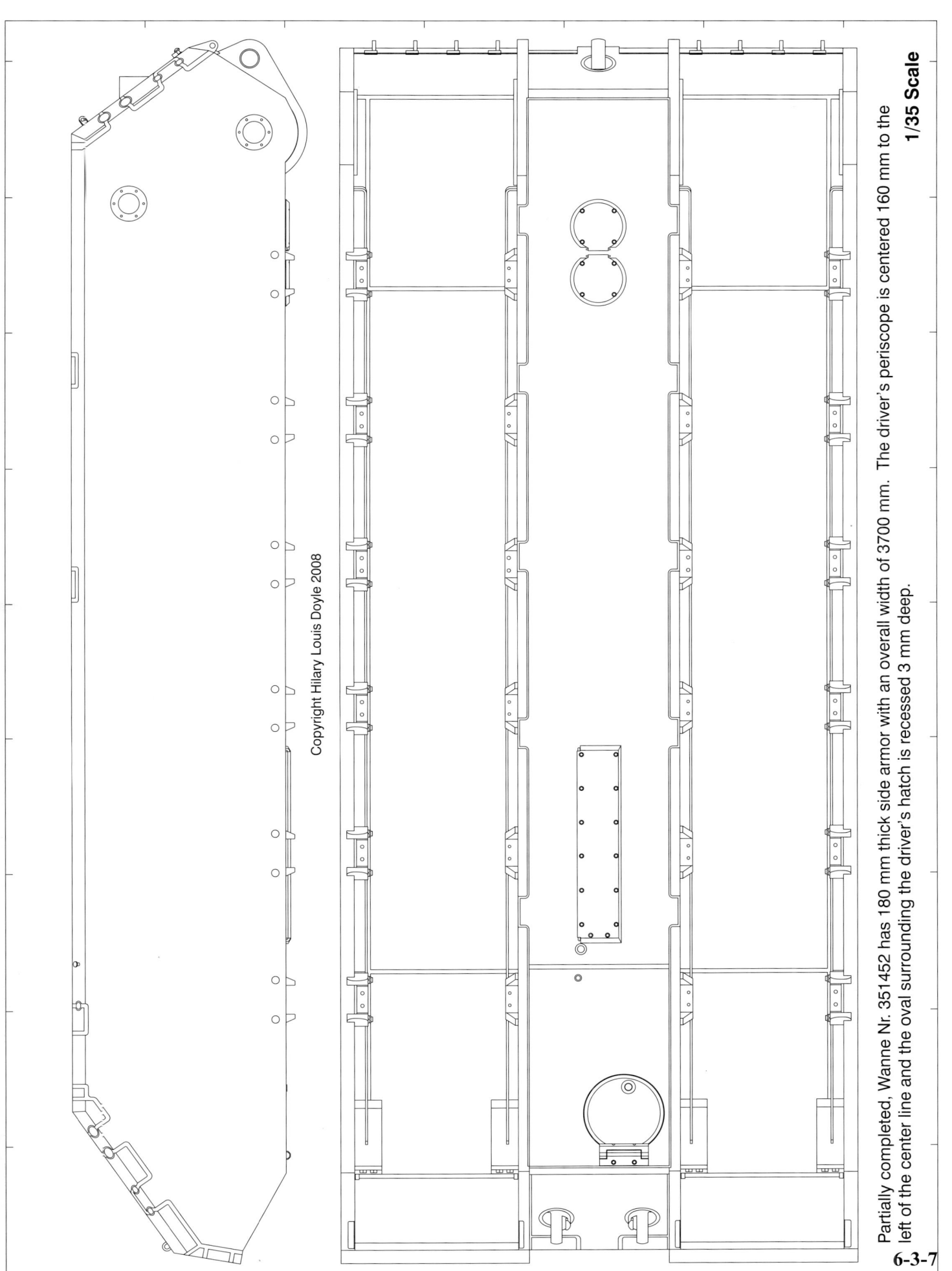

Partially completed, Wanne Nr. 351452 has 180 mm thick side armor with an overall width of 3700 mm. The driver's periscope is centered 160 mm to the left of the center line and the oval surrounding the driver's hatch is recessed 3 mm deep.

1/35 Scale

Copyright Hilary Louis Doyle 2008

6-3-7

## PORSCHE-MAUS DEVELOPMENT

In a meeting at Porsche attended by Krupp representatives on 8 January 1943, Herr Martin described the Porsche proposal for the first **Maus** model that is to be produced:

*The Porsche design has the turret at the rear, so the driver is separated from the rest of the crew. The engine is in the middle. The driver is seated over the **Schaltkasten** (switch panel). He has a seat that can be raised and indirect vision through prismatic periscope. When there isn't any enemy fire, he can open the hatch in the deck. An emergency hatch through the front is therefore not needed. The front armor plate is intact without penetrations. The compartment for the driver and radio operator does not have a machinegun.*

*The electrical generator is located under the turret with the electric motors behind it on each side connected to the rear drive. Armor thicknesses for the hull are: belly and deck 100 mm at the front and 50 mm at the rear. side walls 100 + 80 mm above and 100 mm below, front and rear plates 200 mm, and pannier above tracts 40 mm. The side walls are vertical (not sloped). It still has to be determined if it is more advantageous to make the side walls out of two plates (the first 100 mm plate from top to bottom and a second 80 mm plate at the top, or out of a single 180 mm thick plate that is milled out to create a 100 mm thick lower section. The double plates must be interlocked. Bolting them together is not allowed, because bolt heads can't protrude on the outside.*

*Ventilation air will be drawn in at the front in the middle on top and blown out by fans to the left and right after passing through the radiators. At the front and rear are three part gratings. There are three firewalls and electric cables run along the bottom. The electric motors will be delivered from Siemens and the suspension and tracks from Skoda.*

At a meeting on 18-22 Janaury 1943, **schwenkbare Lukendeckel** (pivoting hatches) were declined by both Wa Pruef 6 and Porsche because when released the lids are quickly pushed open by the springs.

The full-scale wooden model was presented at a Panzer-Kommission meeting in Stuttgart on 21 January 1943 attended by Gen.Lt.Ritter von Radlmaier (Panzerkommission) Oberstlt. Holzhauer, Oberst von Wilke, Oberstlt. Crohn, Oberbaurat Kniepkamp (Wa Pruef 6), Dir. Freyberg (Alkett), Direcktor Dr. Pohl (Skoda), Direcktor Dorn (Krupp) Professer Dr. Porsche, Porsche jun., and Chefkonstrukteur Rabe (Porsche KG). Improvements that were discussed included: *Oberbaurat Kniepkamp reported that a **neue Laufkette** (new track) was being developed by Wa Pruef 6 that has a significant advantage over the Porsche proposal mainly because it weighed 30% less. The **Wannen-Aussigluke** (crew hatch in the hull) is to be enlarged to a minimum 500 mm diameter. A drawing for a neue MG-Kuppel located next to the Aussteigluke was requested from Daimler-Benz. A 100 mm thick sloped plate was added as a **Kettenschutz** (track guard).*

At a meeting with Hdl. Saur (R.M.f.B.u.M.) on 17 February 1943, Prof. Porsche was asked to initiate the production of the Daimler-Benz **Flugmotor DB 603** as a **Spezialausfuehrung** (special model) rated at 1200 metric horsepower with fuel injection.

On 2 April 1943, Porsche recorded that the total weight for the **Maus** had increased about 10 metric tons above the 179.3 metric tons (calculated 1Jan43) due to increased weight of some components (1.5 tons to the turret and 2.2 tons to the chassis because of a plus tolerance on the armor thickness of 3%), additional ammunition stowage (1.46 tons), and a new component, the **Flammenwerferanlage** (flamethrower system) weighing 4.9 tons.

After the full-scale wooden model of the **Maus** was shown to Hitler on 13 May 1943, as discussed in Hitler's conferences with Speer on 13-15 May 1943, Item 23 - *The amount of 12.8 cm ammunition in the **Maus** must be increased from 50 rounds to 80 rounds. There are no objections if this results in reducing the 7.5 cm ammunition from the previously planned 200 rounds down to 100 rounds.*

On 10 June 1943, Wa Pruef 6 informed Alkett: *Additionally, the six **Pz.Kpfw.Maus Versuchsfahrzeuge** will receive a **Gasschutzanlage** (poison gas protection system) that will be delivered ready to install from the Draeger-Werk, Lubeck (from a Wa Pruef 9 contract).*

Additional details on the development history of chassis components (armor hull, new suspension, armor plate, and flamethrower) are presented in the following sections.

## MAUS WANNE (Armor Hull)

On 18 January 1943, a decision had been made to create the side walls out of a single 180 mm plate that is milled out to 100 mm thick at the bottom. Krupp's proposal to mill out the lower section to 120 mm thick with a 60 mm inner wall was not acceptable because the wheel base could not be altered. The inner 80 mm thick walls were to be made out of softer steel.

On 17 March 1943, Krupp sent a letter to Wa Pruef revisiting the decision to make the hull sides out of a single 180 mm thick plate. *Krupp had reservations about milling out about 4.5 tons of steel from each side wall to create a lower section that was 100 mm thick, and proposed that the side walls be constructed by an alternative method. The alternatives suggested were:*
*a. A narrower 80 mm thick plate on a wider 100 mm thick plate. However, due to inexact surfaces, they won't fit close together. The space must be filled with **Zementkitt** or a similar substance. Securing both plates together in a manner which will withstand hits was a special problem.*

*Krupp and Wa Pruef Festung were currently developing and testing studs that won't fly off or be shot into the interior when hit. Krupp didn't pursue this design because 20 to 25 holes and studs were needed for each side wall. A simple solution would be to replace the studs with rivets, but the heads will fly off inside when hit.*
*b. Space both plates 30 to 40 mm apart. This was not acceptable because it would increase the width or take up needed space on the inside. Other than redesigning the hull again, the same disadvantages existed as above.*
*c. Create the side wall out of one piece. This is absolutely the sturdiest type. In addition, it was their current experience that a single plate has better resistance to penetration by large caliber shells than double plates adding to the same thickness. In addition, bolting the plates together would take more work hours than milling out the lower section. Therefore, in spite of the waste of steel, Krupp was in favor of milling out the single plate.*

At a meeting in Stuttgart on 23 March 1943 attended by Oberbaurat Rau (armor specialist in Wa Pruef 6) and Dir. Dorn from Krupp, the matter of the armor hull side walls was again discussed: *All present were unanimous in their opinion that the work expended to mill out the bottom section of 180 mm thick plates was so considerable that another solution must be found. It was decided that because of the production schedule the first* **Maus** *would not be changed, but for the future other simpler methods were to be examined and tested on a firing range. Other solutions suggested were: A single 100 mm plate with the upper 80 mm plate spaced 10 mm apart but not fastened together. A single 180 mm plate for the entire height with a redesigned method of supporting the* **Laufwerk** *(suspension and roadwheels) instead of the 80 mm thick inner hull plate.*

On 2 April 1943, Herr Martin informed Krupp that Porsche intended to outfit the **Maus** with a **neue Laufwerk** (new suspension). The lower section of the outer side wall would need to be only 60 mm thick and could be welded to the upper side wall that remained 180 mm thick. On 6 April 1943, Krupp resported that this **geteilte Ausfuehrung** for the outer side wall (180 mm upper section/60 mm lower section) with a 120 mm thick inner wall could be introduced starting with the **7.Wanne** in the mass production series of 120.

At a meeting between Porsche and Krupp on 26 May 1943: *A decision was made that the* **Verlademass** *(rail loading width of 3700 mm) couldn't be exceeded. As a result, the width of the hull must be changed to include the plus tolerance of 3% for the armor plates. The armor plates for the first four* **Wanne** *were completed and plates were being prepared for another nine* **Wanne**. **Wanne Nr.1 - 4** *are to be welded together without any changes. The maximum width of these four* **Wanne** *is not to exceed 3715 mm. The outer side walls for* **Wanne Nr.5 - 13** *are to be milled down to 170 or 90 mm thick. Starting with* **Wanne Nr.14** *the side plates are to be rolled 170 mm thick. Starting with the* **Wanne Nr.7**, *the front and rear plates, the front track guard, and the forward deck are to be reduced 5 mm in width with a -1 mm tolerance.*

*Porsche also informed Krupp of another modification. The* **Traegerstuetzen** *(supports) for the suspension have been changed from welding to being bolted to the outer sidewalls and 12 holes need to be bored in the outer side walls.*

On 26 June 1943, Porsche informed Krupp: *To reduce the time it takes to rework welded together hulls, which will result in schedule delays, the outside of the side walls for* **Wanne Nr.3 and 4** *are to be milled down to 170 mm thick.*

The following changes were made to milling the **Wanne**, as incorporated into drawing 205.43.04.1 dated 2 June 1943: *The* **Leitradlager** *(idler mounting) will be milled starting with* **Wanne Nr. 2**. *The Leitradlager is already welded into* **Wanne Nr.1**. *The outer* **Tragerstuetzen** *(suspension supports) are 25 mm shorter than those previously delivered from Skoda. The exit point for measuring the 75 mm diameter holes for mounting the* **Kettentrieb** *(track drive) and* **Elektromotorenlagerung** *(electric motor mounts) are 445.84 plus 232.20 mm apart.*

On 22 July 1943, Krupp was informed of additional work on **Wanne Nr.1**: *The front and rear plates are to be bored to mount the* **Augen** *(eyes) for the* **Abschleppgeraet** *(towing equipment). The square* **Deckel** *(cover) on the belly at the rear is now 250 mm longer and the opening will be extended by flame cutting (starting with* **Wanne Nr.1**). *Holes are to be drilled in the front deck for the* **Funkeroptik** *(radio operator's periscope) and for studs to mount the* **Schalttafel** *(switch panel).*

On 21 August 1943, Krupp was informed that the **Augbolzen** (towing eyes) for the **Abschleppvorrichtung** (towing device) for recovery and towing equipment are to be welded into holes in **Wanne Nr.1** to **4**. Starting with **Wanne Nr.5**, these are changed to **Befestigungsteile mit Rundgewinde** (fasteners with round joints) for recovery and towing equipment.

On 9 October 1943, Herr Martin delivered new drawings 205.53.027 and 028 for **Graetings** that were to be installed starting with **Wanne Nr.2**. Several additional parts had been added to the previous design.

## NEUE LAUFWERK (New Suspension)

On 2 April 1943, Herr Martin informed Krupp that Porsche intended to outfit the **Maus** with a **neue Laufwerk** (new suspension). An explanation of the cause for a **neue Laufwerk** for the **Maus** was recorded on 10 April 1943, as follows: *The production schedule was the primary consideration at the beginning of the* **Maus** *design so that available components were to be adopted, such as the*

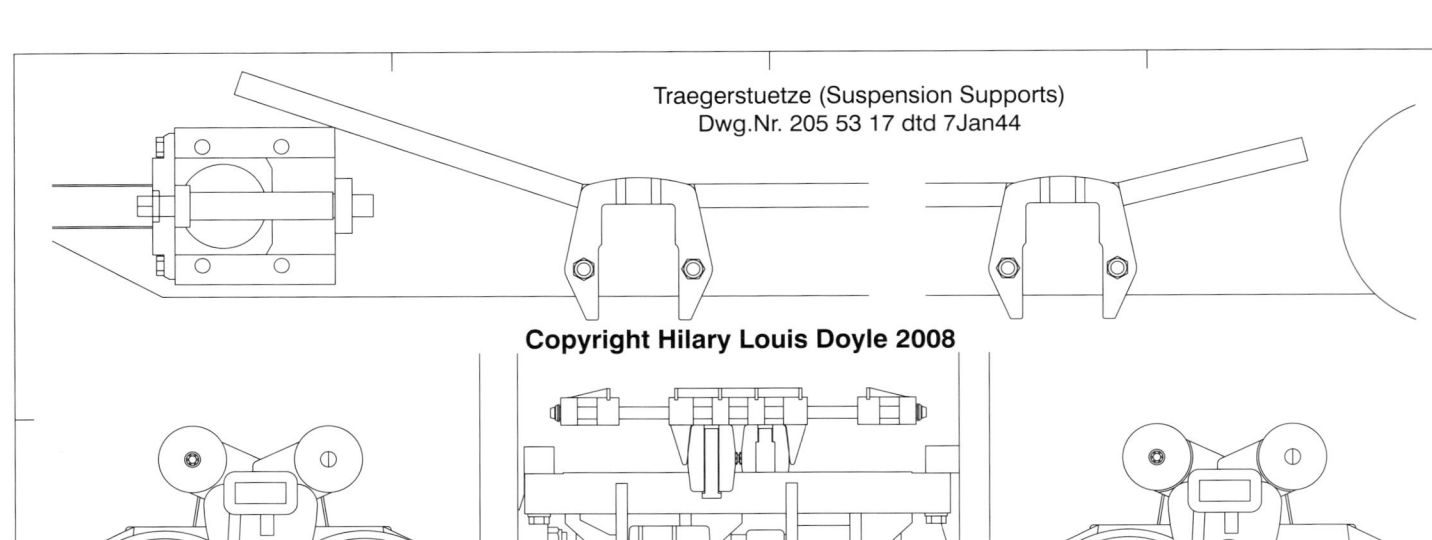

Traegerstuetze (Suspension Supports)
Dwg.Nr. 205 53 17 dtd 7Jan44

**Copyright Hilary Louis Doyle 2008**

Laufwerk (Suspension Assembly) Dwg.Nr. 205.47.00 dtd 1Jul43

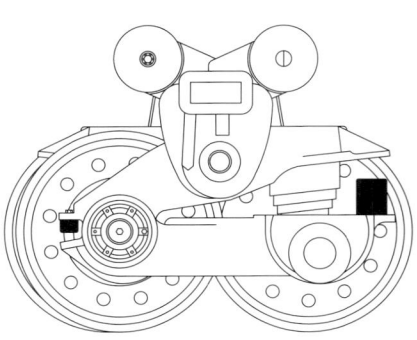

Laufrollen (Roadwheels) with holes drilled to secure rubber rings

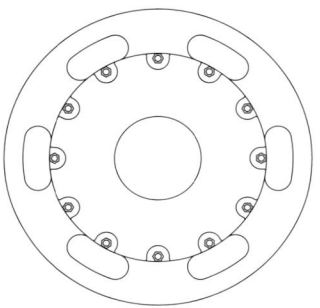

Leitrad (Idler)
Dwg.Nr. 205.28.20
dtd Apr44

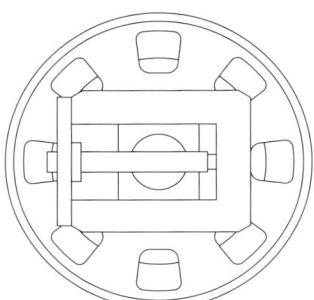

Alternative Leitrad (Idler)
Dwg.Nr. 205.28.15
dtd 27Jul43

Triebrad (Drive Sprocket)
Dwg.Nr. 205.25.01 dtd 10Jun43

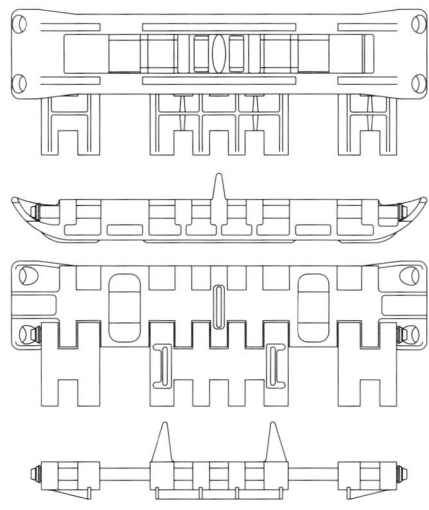

Ketten (Tracks)
Dwg.Nr. 205.48.01 dtd 28May43

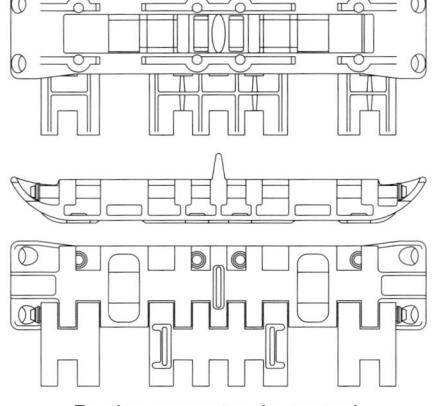

Replacement tracks tested on Maus Versuchs-Fgst. Nr.2

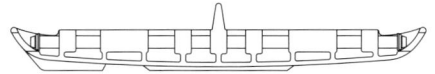

Alternative track design with asymetrical cleats

**1/20 Scale**

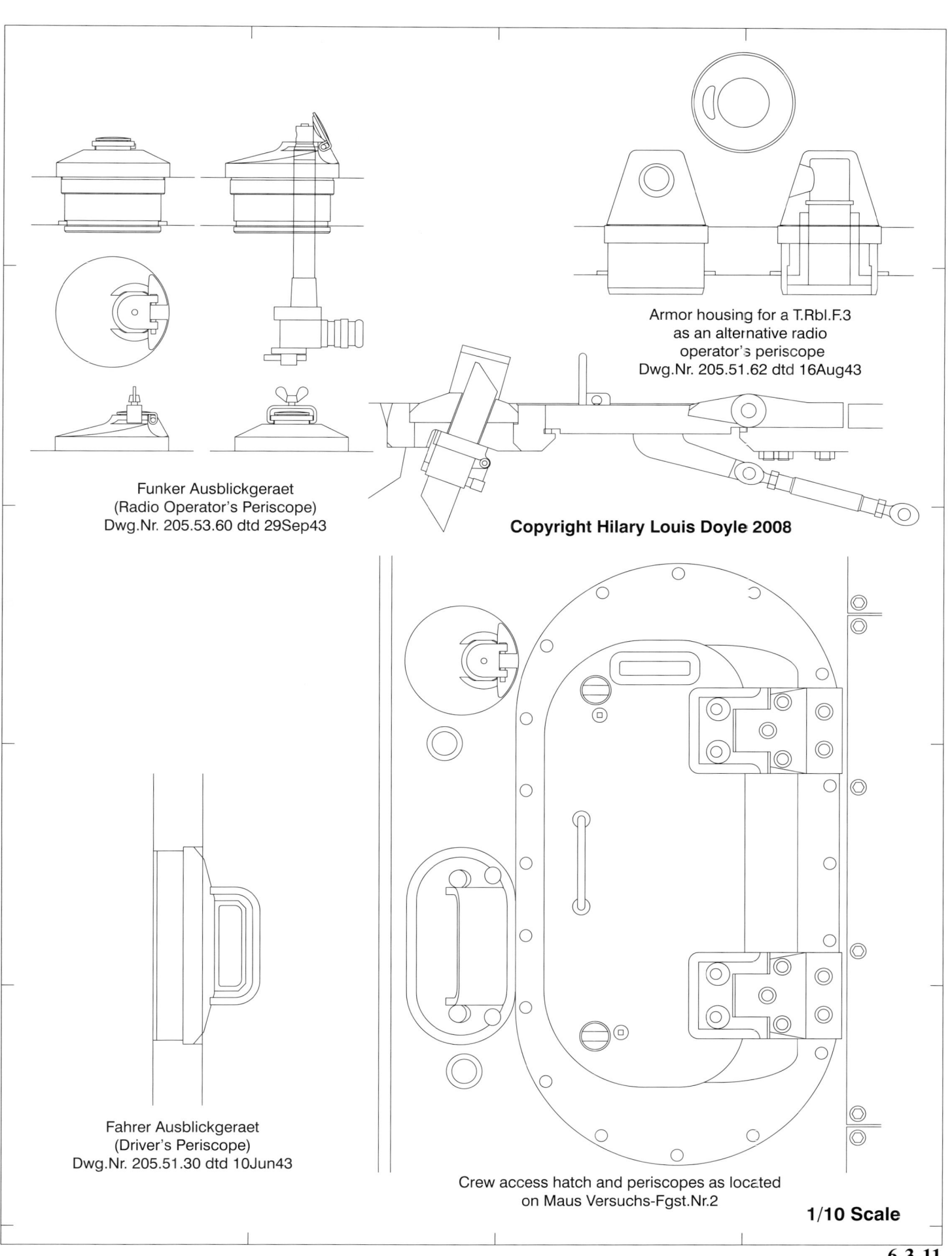

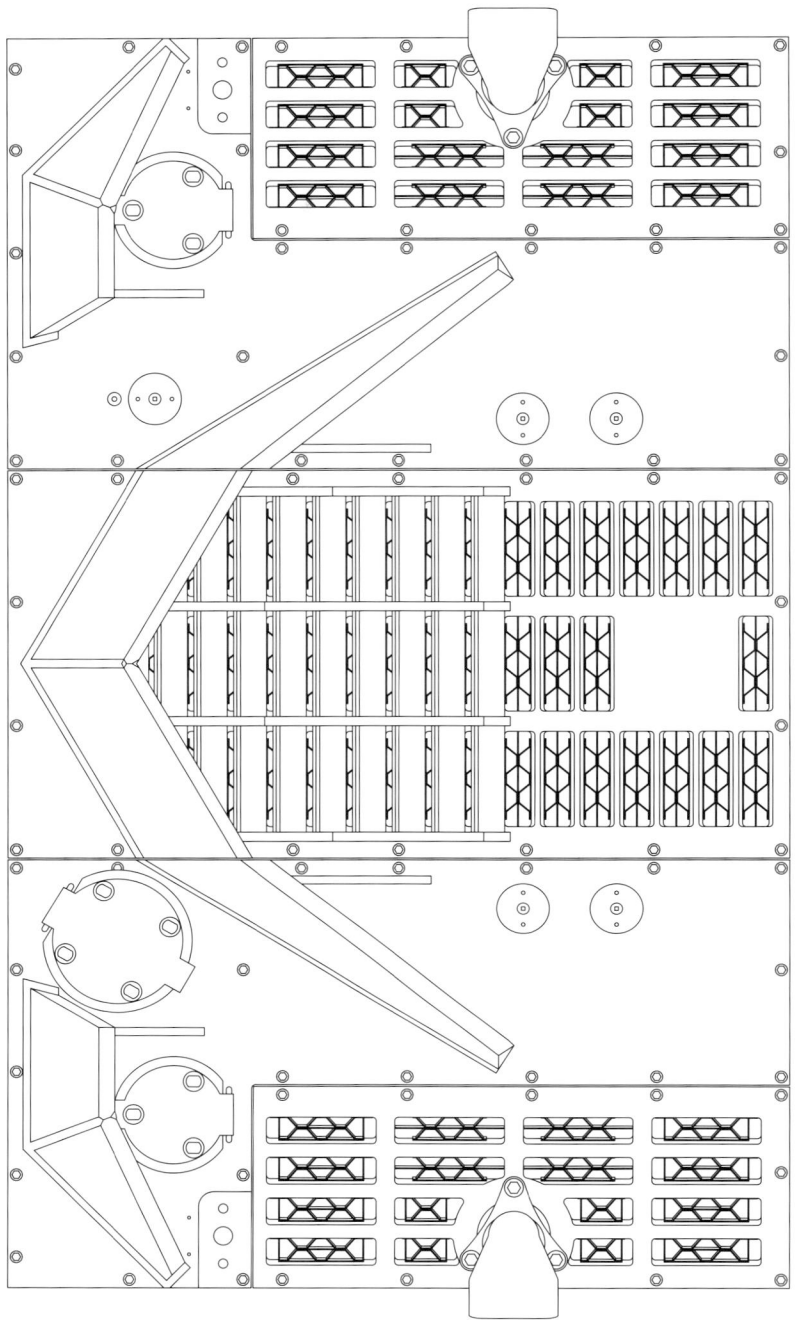
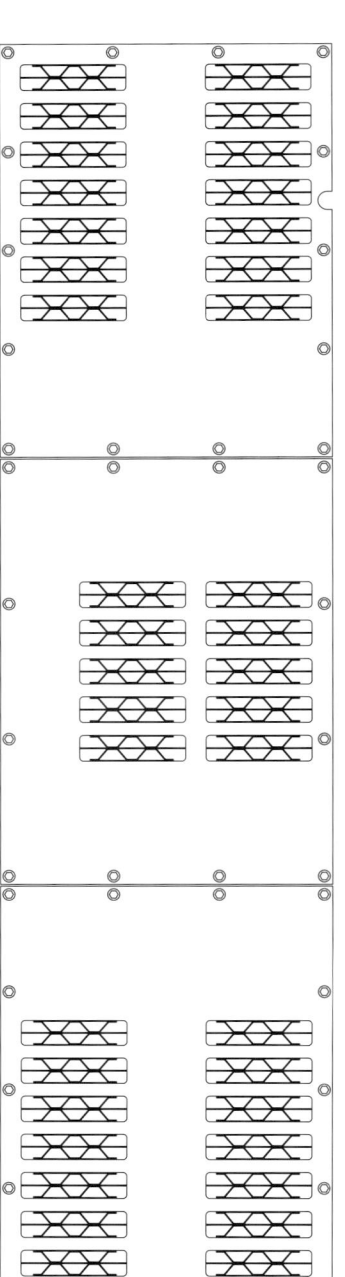
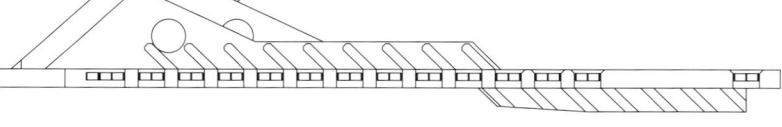
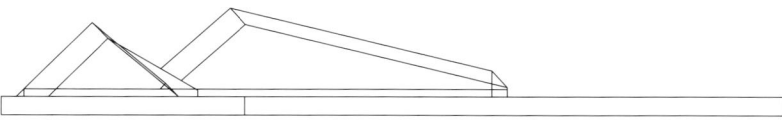
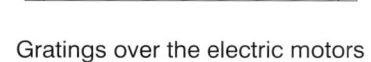

**Copyright Hilary Louis Doyle 2008**

Gratings over the electric motors on Maus Versuchs-Fgst.Nr.2

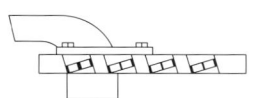

Gratings and shot deflectors over the engine compartment on Maus Versuchs-Fgst.Nr.2

**1/20 Scale**

suspension from the **Tiger**. The original plans for armor protection for this suspension was a 80 mm thick wall that was doubled to 160 mm thick for the upper hull side. At this time the total weight was about 150 tons. Because of strengthened armor and increased turret weight, the **Tiger** suspension had to be strengthened.

Porsche's plans for the design of the side walls with the doubled upper section was turned down by Krupp and, as agreed upon by Krupp, a design was selected that was 180 mm thick with the lower section milled out to 100 mm. Increased weight resulted in thicker **Tragzapfen** (support pins) for the suspension, and holes needed to be drilled through the hull sides to fasten the suspension.

At a meeting in Stuttgart on 23 March 1943, the Hauptausschuss decided that milling the side walls and drilling holes was too much work and too much steel wasted for series production. In addition, in all of Germany, only Krupp had the special milling machine needed to mill out the hull sides. If this milling machine broke down, the entire production series would be halted.

In addition, Oberbaurat Rau did not have experience with penetration of doubled armor plates (in comparison to a single plate of the same total thickness), so that we as the first designer would be taking a major risk that couldn't be reversed. As decided at the meeting on this subject, **Beschussmodelle** (pieces for penetration trials) were to be completed and tested. However, in the interim it was found out the navy had experience that doubled armor plates have only 70% penetration resistance of a single plate of the same total thickness.

The track design, which we showed Oberstlt. Holzheuer and Oberbaurat Kniepkamp in Stuttgart on 21 January 1943, has been modified based on Waffenamt experience. In the interim we learned that the newest track design for the **Panther** and **Tiger** were entirely different from ours.

As a result, we have taken the following actions. A suspension previously developed by Porsche (that has no

**Above: The 1100 mm wide tracks with face cleats that were used on the 2.Maus Fahrgestell. (TTM)**

similar parts to the **Tiger** suspension) was rapidly designed. Our Chefkonstrukteur, Herr Rabe, visited Skoda and explained the new suspension to be produced on time to meet the schedule deadlines. Herr Martin informed Krupp that the difficult side wall milling and drilling holes were not needed for the new suspension.

The advantages of the **neue Laufwerk** are:
1. Another pair of roadwheels can be installed on each side, reducing the load on each.
2. Fastening the suspension in the middle on **Streben** (bracing strips) eliminates the need for drilling holes in the hull sides.
3. Installation of the **Streben** makes it possible to weld the **Kettenschutz** (track guard) onto the bottom of the hull side so that milling work is not needed.
4. Decreased work in producing the suspension.
5. Weight savings of about four tons per vehicle.

In addition, the **neue Laufwerk** can be backfitted into the older hull design without disturbing components installed inside the hull.

On 19 April 1943, Krupp was informed that for securing the new suspension, 24 **Anschlussstuecke** (attachment pieces) supplied by Alkett needed to be welded to the armor hull.

**Above Left and Right: The track and suspension designed by Porsche and produced by Skoda as a replacement for the Porsche longitudinal torsion bar suspension. (TTM)**

## PANZERUNG (Armor)

On 24 January 1943, OKH/Wa Pruef 6 awarded Fried.Krupp A.G. Abt.AK, Essen contract SS 006-4112/42 for an empty **Turmgehaeuse** (turret body) for **Pz.Kpfw. "Maus"** for firing trials. On 26 January 1943, Krupp received contracts to complete two **Beschussmodelle** (empty hulls for penetration targets) built in accordance with the current **Porsche-Maus** design. **Modell 1** with milled bevels and **Modell 2** with flame cut bevels were to be completed by 15 April 1943. Herr Rau (Wa Pruef 6) defended his preference for milling instead of flame cutting based on Hitler's declaration that this should be the best and strongest Panzer. Therefore no design was acceptable that still hadn't been tested on the range.

A meeting was held on 4 February 1943 to discuss obtaining **Marineplatten** (navy armor plates) which had been made available to Wa Pruef Fest. *Major Widerholt agreed that navy armor plates could be obtained for the Maus as long as this does not effect the Atlantik-Wall-Programm schedule. If these navy plates were used, they had to be rerolled to pieces that were 2000 x 2300 x 200 mm thick. The best navy armor could be obtained for the Maus because average armor was good enough for construction bunkers.*

The results of **Beschussversuch** (firing trials) for the **Maus** in Hillersleben were reported on 22 June 1943. The diameter of the **Verbindungsbolzen** (connecting pins) was to be reduced from 100 to 80 mm diameter so that the armor plate would be weakened as little as possible. The conclusion was that the number of **Bolzen** couldn't be reduced.

## FLAMMENWERFER

As initiated by Wa Pruef 6 on 16 December 1942, Hermann Koebe Feuerwehr-Geraete-Fabrik sent a drawing of the **Gross-Flammenwerfer** installed in the **Pz.Kpfw.III** to Krupp on 28 December 1942. It used a pumping system running at 3000 rpm driven by a 30 horsepower two-stroke 1100 cc engine. A larger engine would be needed in the **Maus** to achieve a longer range.

On 23 January 1943, Krupp asked Koebe if they had already designed a **Flammenwerfer** with a range of 150 to 200 meters. Koebe replied that a flamethower with a 22 mm diameter nozzle achieving a range of 100 meters had used 33 liters per second, while a 14 mm nozzle with a range of 50 to 60 meters used 7 to 8 liters per second. Koebe noted that ranges of 150 to 200 meters could hardly be reached.

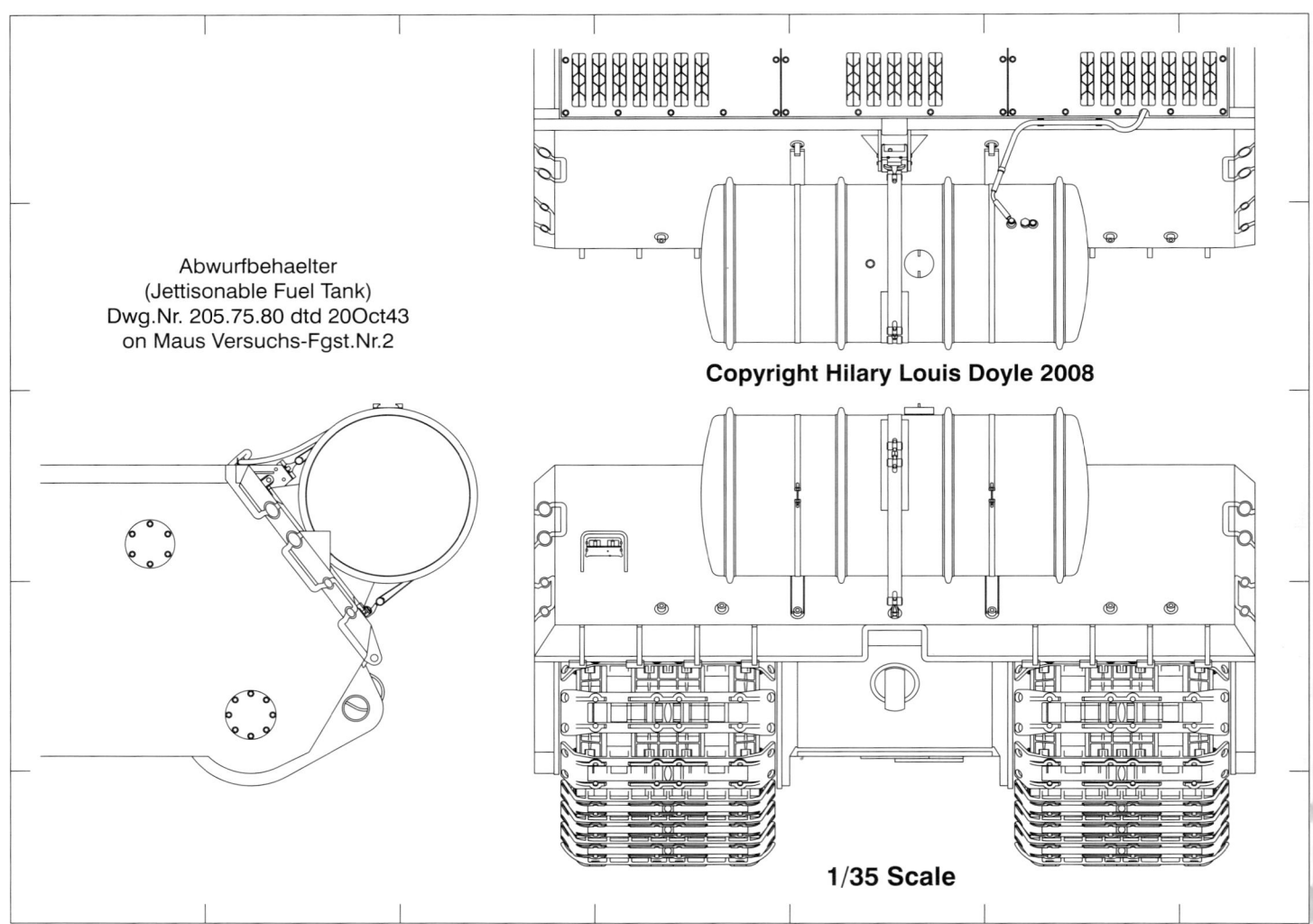

Abwurfbehaelter
(Jettisonable Fuel Tank)
Dwg.Nr. 205.75.80 dtd 20Oct43
on Maus Versuchs-Fgst.Nr.2

Copyright Hilary Louis Doyle 2008

1/35 Scale

Shown a full-scale dummy and three different wooden models of **Spritzkoepfen** (spray heads) on 1 March 1943, Wa Pruef 5 and 6, Wegmann, and Porsche decided on two **Spritzkoepfe Ausfuehrung III** (drawing 6170-3) from Wegmann that were to be mounted in a **Sockel** (socket) on the right and left rear of the **Maus**. They turned down the idea of mounting two **Spritzkoepfe** on the front because (1) if the **Spritzkopf** was shot off the fuel tanks would be endangered by burning **Flammoel**, (2) the chance of the enemy firing into the hole left by the shot-off **Spritzkopf** was too high, (3) the amount of stored **Flammoel** would be reduced by 100 liters, (4) the **Flamm-oel** tanks would need to be a very convoluted shape, (5) the danger of pulling burning **Flammoel** into the air intake gratings in headwinds, and (6) the major difficulty in locating the pressurized lines and aiming device. As directed by the tank commander, the **Spritzkoepfe** mounted at the rear were to be aimed by the radio operator using a remote controlled electrical motor from Himmelwerke. A device installed by the radio operator would show the position of the **Spritzkoepfe**. The **Flammenwerfer** were to have 12 to 14 mm nozzles and a pressure of 14 atmospheres supplied by a pump. The **Flammenwerfer** for the first **Maus** were to be delivered to Alkett by 5 August 1943, followed by a second on 10 October, and the rest spaced three weeks apart. The complete **Flammenwerfer** system weighing 4.9 metric tons consisted of a 1000 liter flame oil tank, a pump with motor, two armored **Spritzkoepfe**, and two control consoles.

Further design considerations were discussed by Oberstlt. Crohn (Wa Pruef 6), Krupp, Wegmann, and Porsche on 6 April 1943. Wegmann was asked to redesign the **Flammenwerfer** installation based on: *The vertical wall of the **Sockel** must be 150 mm thick. Because of the center of gravity of the entire vehicle, it is absolutely necessary to reduce the weight of the **Flammkopf** (now 4 tons without the **Sockel**) down to 2 tons by abandoning the idea of a heavily armored **Flammkopf**. The new **Flammkopf** was to have sloped sides 30 mm thick with 14.5 mm armor on top and below. Due to the reduced armor protection, the **Flammkopf** must be redesigned so that if hit, there wasn't any chance of burning **Flammoel** entering the vehicle.*

As decided on 10 April 1943, Oberbaurat Rau (Wa Pruef 6) decided that the two protrusions to be added onto the upper hull rear for the **Flammenwerfer** would be in the form of horseshoes with 150 mm thick walls, have an inner radius of 120 mm, and be interlocked with the upper hull plate.

After the full-scale wooden model of the **Maus** was shown to Hitler and representatives from In 6 on 14 May 1943, it was reported to Krupp that the decision had been made to drop the **Flammenwerfer** system, and the associated openings for it in the hull rear could be eliminated.

**Above Right and Left:** The full-scale wooden model of the Porsche-Maus on display for Hitler and entourage on 14 May 1943 still had a commander's cupola on the roof. In 6 decided that the Flammenwerfer (flame thrower) system (mounted on the right and left upper hull rear) would be dropped. (BSB)

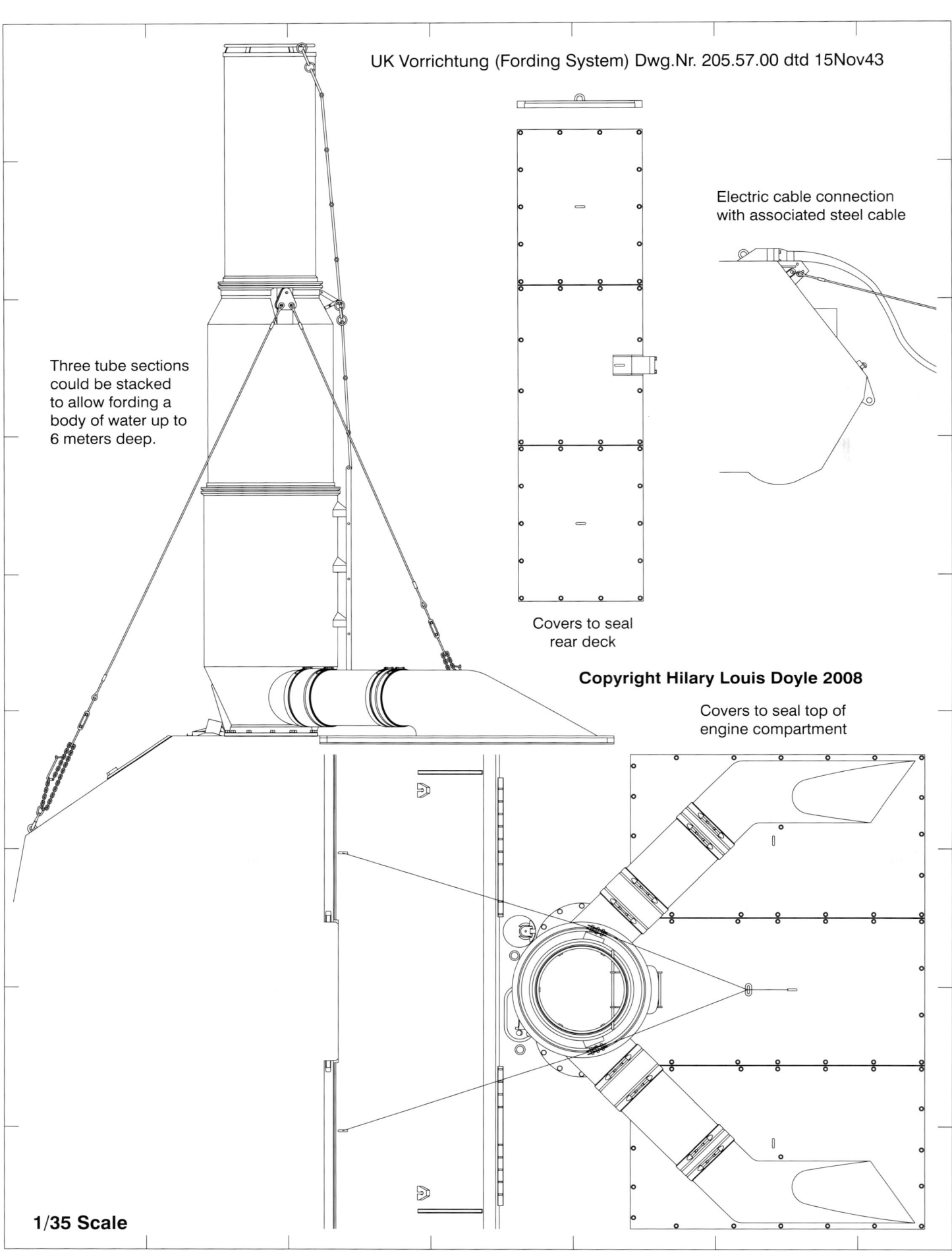

## MAUS TURM DEVELOPMENT BY KRUPP

On 12 January 1943, Obering. Dorn (Krupp) met with Oberstlt. Crohn (Wa Pruef 6) in Berlin to discuss the **Maus Turm 12.8 cm**. Krupp turned over proposals for the individual turret components, which were fundamentally agreed upon. The following changes were requested:

1. *The gun should be a **12.8 cm Kw.K. L/55** with a penetration ability of 250 mm/1000 m/60° with a **Sondergeschoss** (special projectile). The **7.5 cm Kw.K. L/36** remains unchanged, with the cartridge taken over from the L/24 unchanged. The carriage and turret are to be dimensioned so that a **15 cm Kw.K. L/38** can also be installed.*
2. *The machinegun is deleted. In its place try to install a **2 cm Flak** in a horizontal fixed mount without movement for aiming.*
3. ***Flammenwerfer** is to be mounted in the chassis and not the turret.*
4. *A **Nebelwurfgeraet** with 360° traverse is to be mounted in the turret roof.*
5. *A **Kippspiegel-Zielfernrohr** (periscopic gunsight) is planned for the gunner. All other sights such as a **Pak-Zielfernrohr** and **Rbl.Fernrohr** are dropped. This significantly simplifies the linkage for the gunsight.*
6. *The possibility of installing a newly developed **E-Messer** (range finder) is to be investigated.*
7. *The turret is ventilated from the chassis by 5 m³/min of air blown in that exhausts around the gun mantle and turret ring. A capped opening is planned for ejecting spent cartridges in which later, if needed, a ventilation fan can be installed.*
8. *The **Kommandantenkuppel** is to be strengthened to match the turret armor.*
9. *The **Einstiegklappe** (crew hatch) is to be increased from 50 to 60 mm thick.*
10. *Penetrations for the **Antennen** leads in the hull deck and the antenna are to be laid over by the traverse mechanism.*
11. *Wa Pruef 6 wants a small gasoline/electrical generator.*
12. *All of the side **Ausblichluken** (vision ports) and spent cartridge ejection ports are dropped. **Schwenkspiegel** (periscopes) are planned to be mounted in the roof of the turret so that the gunner can observe all round.*

On 24 February 1943, Porsche wrote Krupp concerning the weight of the **Turm Typ 205 ("Maus")**: The weight of the complete **Turm** including base ring and ammunition was originally 49.5 metric ton and later increased by Krupp to 51 tons. In the interim Prof. Porsche has been informed that in no circumstances can the total weight of the vehicle exceed 180 tons. This can be possible only if the turret weight is limited to a maximum of 50 tons.

On 16 April 1943, Wa Pruef 6 met with Dr. Bankwitz (Krupp) to determine additional details: *Contrary to the previous decision that there wouldn't be any **MP-Luken** (machine pistol ports) in the side walls, now an **MP-Luke** is to be installed in both side walls. Krupp is to obtain a drawing of an **MP-Luke** with a small **Kugelblende** (ball mount) from Daimler-Benz.*

*Installation of the rotating and tiltable **Prismenspiegels** (periscopes) based on M.A.N. or Henschel models in the turret roof has encountered difficulties because they were designed for thin armor plate and don't work in thick plate. Therefore it is recommended that two **Prismenspiegel** (designed by Porsche for the driver) be located in*

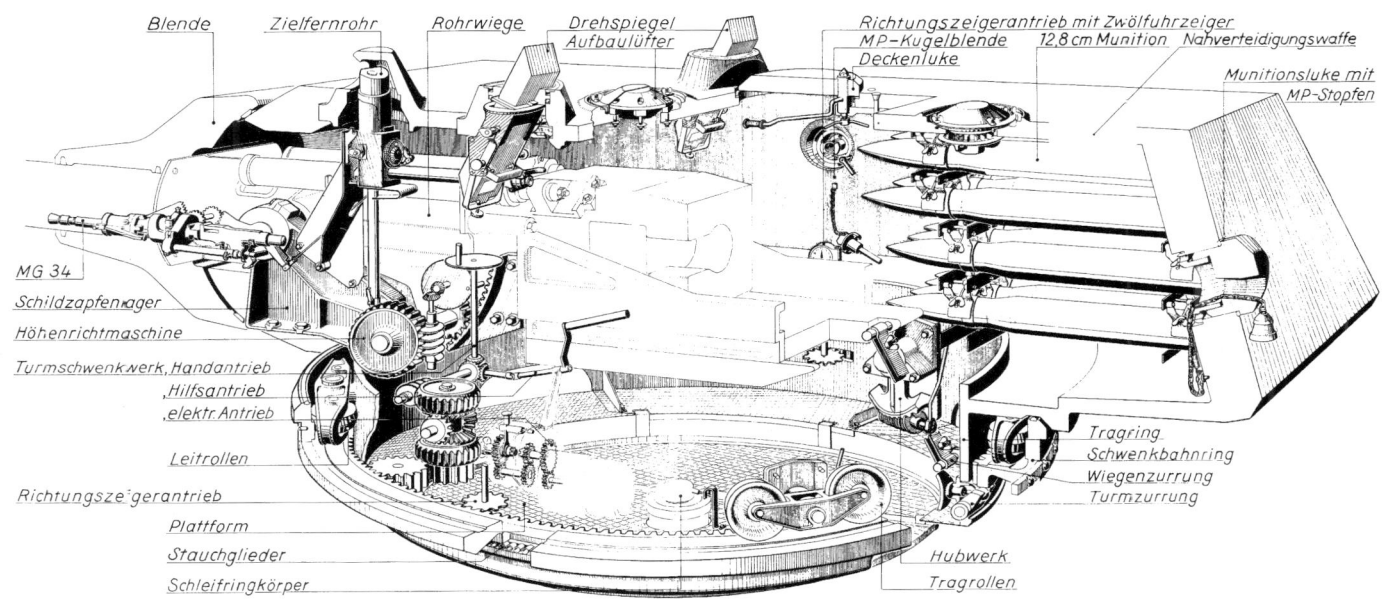

**Above: A sketch made before final modifications. The left Drehspiegel (traversing periscope) was replaced by an Entfernungsmesser (range finder), the right Drehspiegel was replaced with a T.Rbl.F.3 (observation periscope), and stowage for two-piece 12.8 cm ammunition replaced single-piece rounds.**

*the turret roof; one to the right and left for the loaders.*

Three openings for **Montageoesen** (lifting rings) with caps are planned for the turret roof.

Wa Pruef 6 agrees to installation of the **Deckenluefter** (roof vent fan). Whether a second **Luefter** nearer to the **Nebelwurfgeraet** is needed can be determined by trials with the first **Luefter**. Space is available.

As of 16 April 1943, the drawing numbers for the **Maus Turm** components were:

| | |
|---|---|
| 21-4401.001 | senkrechtes Geschuetz |
| 21-4401.006 | Kommandantenkuppel |
| 21-4401.008 | Kippspiegel |
| 21-4401.016 | UK-Vorrichtung |
| 21-4401.019 | Antrieb der Zieleinrichtung |
| 21-4401.025 | Richtungszeiger |
| 21-4401.030 | Kommandantensitz |
| 21-4401.033 | Munitionslagerung 7.5 cm |
| 21-4401.035 | Zubehoerlagerung |

The following components were dropped:

| | |
|---|---|
| 21-4401.015 | Richtantrieb fuer M.G.42 |
| 21-4401.022 | Rohrausblasevorrichtung |

Krupp met with Wa Pruef 6 to discuss the layout inside the turret on 29 April 1943:

*Room for the **Kommandant** is very restricted. He is hindered by the breech of the **12.8 cm Kw.K.** and also the recoil guard on the **7.5 cm Kw.K.** He can stand only while traveling and during combat. When the breech of the 12.8 cm is opened, he must turn his body 90 degrees to the left to avoid it. The space available can first be exactly determined when the full-scale wooden model is completed (rebuilt in about four weeks as a shell).*

*Together with the new vision devices for the commander, Krupp proposes to relocate the hatch from its current position over the breech of the **7.5 cm Kw.K.** toward the rear over a loader, so that it is behind the commander.*

*The lid for the **Kriechkanal** (crawl space) may not be locked from inside the turret. It has to be installed so that the driver and radio operator can open the lid themselves from inside the crawl space. Previously this opening could be used only when the turret was at 12 o'clock. Krupp had asked Porsche not to install further openings in the platform base for rigidity reasons.*

Herr Rabe (Porsche) telephoned Krupp on 8 May 1943, stating: *Porsche and the Waffenamt had decided to install a machinegun aimed to the left through the **Munitionierungsluke** (ammunition port). Krupp informed him that such decisions shouldn't be made without contacting the design firm. In this case, the desired machinegun can't be located at the **Munitionierungsluke**, because this space is needed for loading the **12.8 cm Kanone**.*

After the wooden model of the **Maus** was shown to Hitler et al. on 14 May 1943: *Porsche reported that there was strong opposition to the rounded front of the turret. A shape is to be found in which hits aren't deflected under the turret. A shape was discussed in which the curve below mid-turret was bent forward instead of backward. In addition, an all-round traversable **Fliegerabwehrkuppel** with a 3.7 cm gun was requested. A machinegun should also be installed in the front and rear of the turret.*

The wooden model of the **Maus Turm** was shown at a meeting on 7 July 1943 in Essen attended by representatives from In 6, Wa Pruef 6, Gen.Insp.d.Pz.Tr. Wa Pruef 8, Zeiss, and Krupp. It was almost complete and only lacked the travel locks for the gun, mount for the machinegun, and stowage for the **7.5 cm Granaten**.

*The position of the commander at the **Drehspiegel** remained unchanged from the previous inspection on 5 June. If the commander needs more space in the front, this is possible only with a recoilless **7.5 cm Kanone**, but this would gain a maximum of only 70 mm. The commander will not be hindered by the loader loading the **7.5 cm Kw.K.***

*The loader on the right is very significantly hindered in loading when the breech of the **12.8 cm Kw.K.** is open. In 6 thought that it wasn't possible for the loader on the right to have free movement. He isn't hindered when loading the **7.5 cm Kw.K.** but would be strongly hindered by a **Entfernungsmesser** (range finder) or **Scherenfernrohr** (scissors periscope) mounted in the open hatch above the loader.*

*If the commander has the task of using the rangefinder, he will exchange places with the loader on the right. But while he is range finding, the **12.8 cm Kw.K.** can't be loaded or fired and he won't be protected in the opened hatch.*

*The loader on the left (as well as the loader on the right), can't work the clamps in the ammunition racks. The left loader is also strongly hindered by the tight space between the projectile points and the **12.8 Kw.K.** breech. Both ammunition racks in the turret rear take away room to sit. If the loader sat here, he could loosen the clamps on the cartridges and it would be a significant improvement in his loading movements. By changing the position for the left loader, the **Munitionstransportanlage** (ammo transport system) in its present form is superfluous, and dropping it results in significant simplification. This loader can also operate the traversable **Nebelwurfgeraet**. Its ammunition is close at hand.*

*During combat and while traveling, the gunner's left leg is stretched out to the front and his right leg is bent back underneath. This is unacceptable to both the Gen. Insp. and In 6. By further investigating better space for the gunner, it turns out that by moving the left loader to the rear, more space was gained for the gunner to his rear so that he can lean up to the gun sight and also operate his traverse and elevation wheels. Relocating the seat must result in gunners with long thighs having a comfortable place with both legs inside the rotating platform.*

The gunner is to service the machinegun. To prevent frequently changing the cartridge bags, Krupp is to design a large ammo bin for two rows of belts.

Wa Pruef 6 is to send Krupp information on installation of the **Fliegerbeschussgeraete 42** (mount for an anti-aircraft machinegun) on the turret.

The commander must have the same **Richtungsanzeiger** (azimuth indicator) as the gunner, marked from 1 to 12 o'clock.

At a meeting in Berlin on 25 September 1943, Wa Pruef 6 wanted to install the **M.G.42** with the spring cushioned mount from the **Tiger**. Krupp reported that they were working on mounting the **M.G.34** on the left trunnion cap.

The proposal to install the **Nebelwerfer** based on the Daimler-Benz design in the **Maus** turret was turned down on 12 October 1943, and the **Nahverteidigungswaffe** was to be installed as shown in drawing 21 B 7658. The lower retaining flange on the **Nahverteidigungswaffe** was to be enlarged so that the bolts are located outside the traversed zone.

Krupp reported on 14 October 1943 that when the 220 mm front armor plate is bent into a curve, the thickness on the front is reduced to 205 mm.

Additional details on the development history of turret components (armament, commander's cupola and periscopes, anti-aircraft machine gun, and range finder) are presented in the following sections.

## WAFFEN (Armament)

At a meeting at Porsche in Zufferhausen with Wa Pruef and Krupp on 23 December 1942, it was decided.: *The 7.5 cm Kw.K. must be L/31 calibers long instead of L/24 because the muzzle pressure will be unfavorable for the air intake openings for the engine and cooling system.* A length of L/32 to L/33 was requested on 28 December 1942 so that the muzzle gases would not enter the **Graetings**. About 20 rounds of 7.5 cm ammunition could be stowed in the turret.

Krupp sent a telegram on 26 January 1943 with the information: *The maximum length of L/55 for the 12.8 cm gun in the Maus Turm is the longest that can be carried. The muzzle brake has already been dropped. A longer barrel would require that the entire turret be redesigned, and probably even the chassis.*

The **Maus** armament was again discussed by Wa Pruef 4 and Krupp on 5 February 1943: *In regard to Hitler's request, the main gun is planned to be a **12.8 cm Kw.K. L/55** or a **15 cm Kw.K. L/38**. The design is to be based on the higher impulse of the 15 cm gun with muzzle brake removed so that the 12.8 cm gun can be used without a muzzle brake. This is favored because the **T.S.-Geschoesse** (sub-caliber discarding sabot shells) can only be fired out of guns without muzzle brakes.*

*A **7.5 cm Kw.K. L/36** is to be mounted in the same carriage as the main gun. This gun length is needed because with an L/24 the muzzle gases would hit the chassis air intake openings. The breech is to be taken over from the **7.5 cm Kw.K. L/24**.*

*Drawing numbers for the guns are **5-1208** for the **12.8 cm Kw.K.**, **5-1531** for the **15 cm Kw.K.** and **5-0776** for the **7.5 cm Kw.K.***

On 13 February 1943, Wa Pruef 4/II awarded Krupp contract SS0004-3253/42 to produce guns for the **Pz.Kpfw.Maus**, including:
3 complete guns (each with a **5-1208** and a **5-0776**)
2 gun tubes with breech **5-1208**
5 gun tubes **5-1208**
2 gun tubes with muzzle brake and breech **5-1531**
3 gun tubes **5-1531**
2 firing stands to be delivered to the commander of the Versuchsplatzes Hillersleben.

**Right:**
The coaxially mounted 7.5 cm Kw.K.44 L/36 also shared the same cast gun mantle as the 12.8 cm Kw.K.44 L/55. (KSM)

At a meeting with the R.M.f.B.u.M. on 17 February 1943 attended by Hdl. Saur, Panzerkommission (Thomale, Dr. Porsche, Dorn), Wa Pruef 6, Alkett, and Woelfert (Krupp), Hdl. Saur again brought up the question whether the **12.8 cm Kw.K. L/70** should be specified. Wa Pruef 6 confirmed that the **L/55** was the correct gun considering the circumstances of installation in the **Kampfwagen**. Hdl. Saur was also informed that using **Triebspiegel ohne Hartkern** (discarding sabot without a tungsten core) didn't add much.

On 23 Feb 1943, Wa Pruef 4 informed Krupp: The guns for the **Maus Turm** are the **12.8 cm K L/55** and the **7.5 cm Kw.K. L/32** lengthened to L/36 by a cylindrical extension. The **15 cm Kw.K. L/38** must be capable of being mounted in place of the **12.8 cm Kw.K. L/55**.

At a meeting with Wa Pruef 4 and Krupp on 24 February 1943 it was decided: *The **Maus** will have a **7.5 cm Kw.K. L/36** with lands and grooves only cut to L/32 because the **Ziehbank** (milling machine) doesn't permit longer cuts. It is not necessery to screw on a four-caliber-long extension piece; the gun tube can be maintained as a single-piece tube.*

In a meeting in Berlin on 4 March 1943 Wa Pruef 1/Pak and Krupp discussed ammunition for the **7.5 cm Kw.K. L/36**: *Only the **HL Granate** (hollow charge shell) is still being used as the standard shell for the **7.5 cm Kw.K. L/24** against both field and armored targets. Because the **HL-Granate** fails to penetrate **Schottenpanzer** (shadow armor), penetration of the **7.5 cm Pzgr.39** should be determined for a range of 100 meters at 30 degrees. Penetration of less than 50 mm is not interesting. If usable penetration is achieved, **Pzgr.39** can be used whose quality isn't good enough for the **7.5 cm Pak 40** and therefore have been sorted out.*

On 15 March 1943, Wa Pruef 1/Div.Art. reported that the following rounds are planned for the **12.8 cm Kw.K.**: *Vollkaliber-Pzgr., Tsp.-Pzgr., HL-Gr., Spgr., Be-Gr., Minen-Gr., Brand-Gr., Nebel-Gr., and a **Leuchtgeschoss**. The specification for the **Spgr.** and **Be-Gr** is penetration of 40 cm thick concrete. The shells are to have a **L'Spur** (tracer) with 3000 meters burning length, and in addition a new requirement for a **L'Spur mit Brandsatz** (incendiary) has been added.*

On 29 April 1943, Wa Pruef 1 intended to test the firing rate of both two-piece and cartridge ammunition. Krupp was to perform this test by firing 15 rounds of both types using a **12.8 cm Flak 40** mounted in a **21 cm Moerser-Lafette** surrounded by a wooden model of the turret.

It was determined on 5 May 1943, that an external travel lock was not needed for either the 12.8 cm or 15 cm gun in the **Turm Typ 205** because the guns are mounted at their center of gravity and there are internal travel locks inside the turret.

Wa Pruef 1 was developing new ammunition for the **12.8 cm Kw.K.**, **Pak**, and **Kanone 43** on 29 June 1943. *Initially **Patronen-Mun.** (single piece) must be produced for the **12.8 cm Kw.K. (Maus) L/55**. A contract was awarded for 300 **Patr.Huelsen** (cartridges), with 100 to be delivered by 15 July 1943. Two-piece ammunition is planned for the **12.8 cm Kan.43 (Rh)** and **Pz.Jg** and the same for future production for the **12.8 cm Kw.K. (Maus)** because difficulties are encountered in the acquisition of **Patr. Huelsen**.*

Wa Pruef 1 awarded Krupp a contract to develop a **Treibspiegel-Geschoss mit H-Kern** for the **12.8 cm Pz.Jaeger K. L/55 (Maus)** on 2 July 1943. With a muzzle velocity of 1260 m/s it had a penetration ability of 245 mm at 1000 meters at 30 degrees. If possible, the sub-caliber core was to be the **8.8 cm Pzgr.40**.

During the period from 16 to 18 November 1943, 54 **12.8 cm Pzgr.43** weighing 28.3 kg and 35 **12.8 cm Pzgr. L/3.8** weighing 28 kg were test fired at Schiessplatz Meppen to select the **12.8 cm Pzgr.** for the **12.8 cm Kw.K.44 "Maus"**.

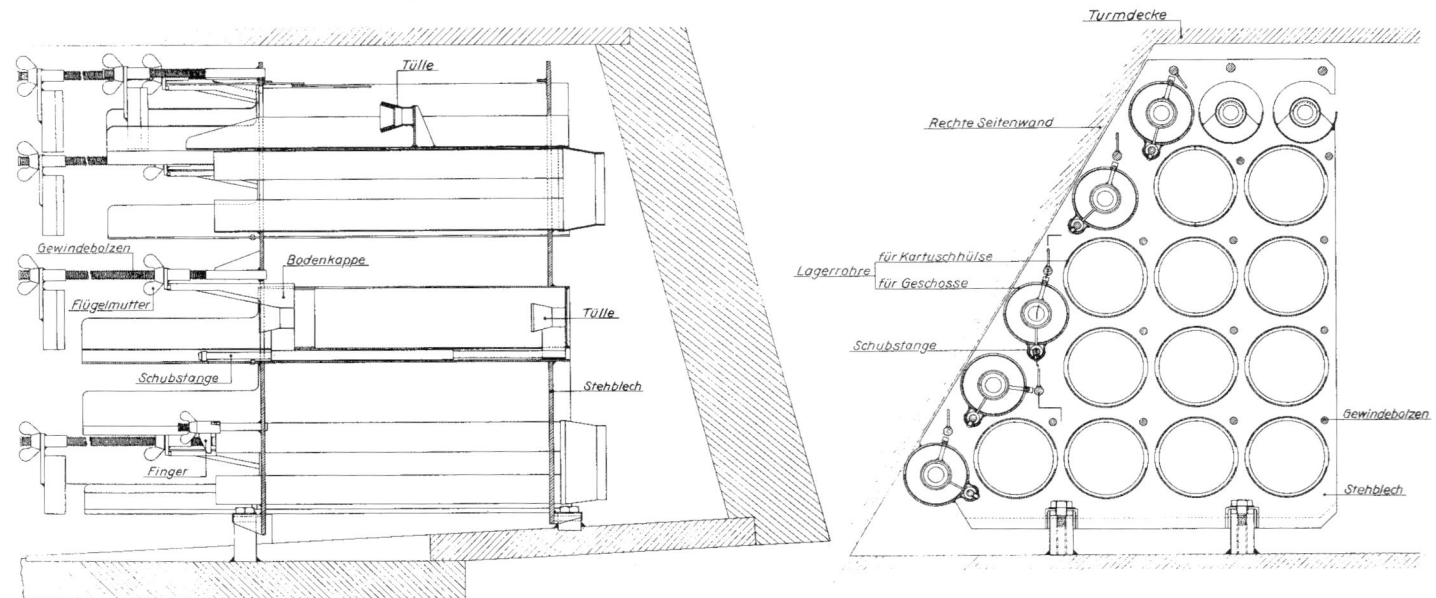

The **7.5 cm Kw.K. L/36** was referred to as the **7.5 cm Kw.K.44 (Maus)** on 6 December 1943.

On 8 June 1944, Wa Pruef 4 stated that they had no interest in completing production of the two complete gun tubes or five gun tubes for the **15 cm Kw.K. L/38**. Contract SS 4915-0004-3253/42 was rescinded.

## COMMANDER'S CUPOLA AND PERISCOPES

On 19 March 1943, Krupp sent a letter to Wa Pruef 6 with drawing Bz 2435 of a **Kommandantenkuppel fuer Maus** attached. *We note that the **Spiegeloptik** (periscopes), **Skalenringanordnung** (azimuth ring), and **Stirnpolster** (cushions) were taken from the **Pz.Kpfw. Tiger** commander's cupola. The cast armor ring for the periscopes also has the same shape on the **Tiger**; however, because of the turret roof thickness, it is 30 mm higher. The **Lukendeckel** (hatch lid) is raised and lowered by a spindle and hand crank, with its weight counterbalanced by a spring.*

Oberstlt. Crohn of Wa Pruef 6 replied on 26 March 1943: *Wa Pruef 6 requests that the following points be re-examined again:*
*1. The edge of the lid sticking out over the cupola has a good chance of being hit, resulting in the lid being easily torn off.*
*2. The pivot axle is located in a zone likely to be hit and the simple welded joint can be easily destroyed by a hit.*
*3. Is operation of the **7.5 cm Kanone** hindered by the location of the pivoting device? Interference with the commander's rear or side may not occur.*
*4. The lid must be opened from outside in order to recover wounded.*

At a meeting in Berlin with Wa Pruef 6 on 29 April 1943 *The drawing Bz 3028 of a **Kommandantenkuppel** can not be accepted because the inner diameter of 250 mm is not sufficient for a head. Instead of this **K-Kuppel**, Krupp proposed a **drehbare Winkelspiegel** (traversable periscope) and **T.Rbl.F.3**. (all-round observation periscope).*

*The commander is to have a 360° **drehbarer Winkelspiegel**, the same as designed by Porsche for the driver in drawing 205.53.014 dated 22Apr43. This periscope must be raised several centimeters so that the commander can see the ground as close as 15 meters in front of the chassis. A small armor collar may be needed on the turret roof.*

*Krupp proposed the installation of the **T.Rbl.F.3** as installed in the **Pz.Sp.Wg. (5 cm) (Sd.Kfz.234)**.*

*Carl Zeiss sent a wooden model of the **T.W.Z.F.1** gunsight for the **Maus** to Krupp on 26 June 1943. The observation opening in the armor guard is cut out for -7 to +23 degrees.*

Krupp sent a telegram to Wa Pruef 8 on 3 August 1943: *The original intention to use the the **Winkelspiegel** 1218 has failed because it is set at a significantly different angle than can be used on the **Maus**, and in addition it must be much longer. The Porsche designed **Fahrerausblick** (driver's lookout) fails because the 180 x 44 Glasblock doesn't fit in the mount for the **Kommandanten-Optik**. The mount can't be enlarged because of the entire layout of the **Maus** turret roof.*

On 2/3 November 1943 at a meeting in Berlin, Wa Pruef 6 and Krupp discussed mounting the **T.Rbl.F 3** in **Maus Turm Nr.1**: *A hole for periscope housing it to be cut into a 200 mm thick armor plate and mounted in the hole previously intended for the **Drehspiegelhaube** (traversable periscope body). As shown on the field of view drawing, vision is worse toward three sides and only better to the right side than the **Drehspiegel**. This is caused by the **Ausblick** of the **Rundblickfernrohr** being only 160 mm above the turret roof, in comparison to 200 mm for the **Drehspiegel**.*

## ENTFERNUNGS-MESSER (Range Finder)

On 3 March 1943, Wa Pruef 6, Zeiss, and Krupp met to discuss an **Entfernungsmesser** (range finder) for the **Maus Turm**: *The following preliminary ideas were discussed:*
*1. **Horizontal-EM 1.73 m** like for the **Tiger**. It isn't possible to take over this design unmodifed because of the location of the guns and the profile of the turret doesn't permit it to be installed.*
*2. **Vertical E.M. 1 m** cannot be decided until a wooden model of the **Maus Turm** is completed. It must be determined if one of the loaders can use the range finder in place of the gunner.*
*3. T-shaped **E.M.** combined with the gunner's sight is only in the conceptual stage. It requires a lozenge-shaped housing on the turret roof about 80 cm wide and 20 cm high. From the start this was viewed as an ideal solution, but it has the disadvantage that the commander's field of view is restricted for about 60 degrees from 12 to 10 o'clock and with a 70 cm base is less accurate for long-range fire. It has the advantage that the gunner himself can find the range and aim at the target.*

When In 6, Wa Pruef 6, Gen.Insp.d.Pz.Tr. Wa Pruef 8, and Zeiss met with Krupp in Essen to view the full-scale wood model of the turret on 7 July 1943: *In 6 requested that the range finder be protected behind armor. This requires a new design. In 6 and Gen.Insp. require that the range finder be used when the hatch is closed. This requirement means that the V-shaped range finder is no longer usable. The only expedient solution appears to be installation of a vertical range finder with the head protruding from a hole in the turret roof. The space for this and its mounting can first be clarified when a wooden model is delivered by Zeiss to Krupp.*

At a meeting at Alkett in Berlin Spandau on 19

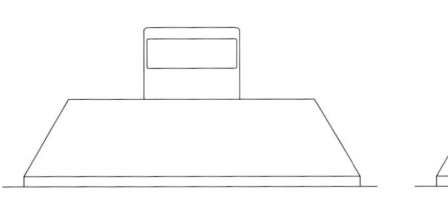

Drehspiegel
(Traversing Periscope)
Dwg.Nr. 021 B 4401.008
(replaced by T.Rbl.F.3)

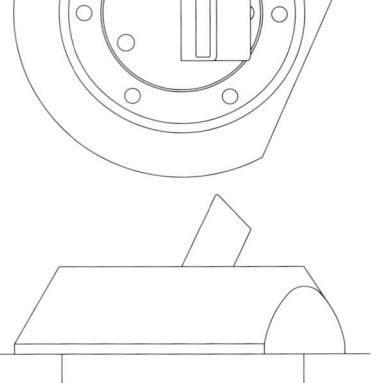

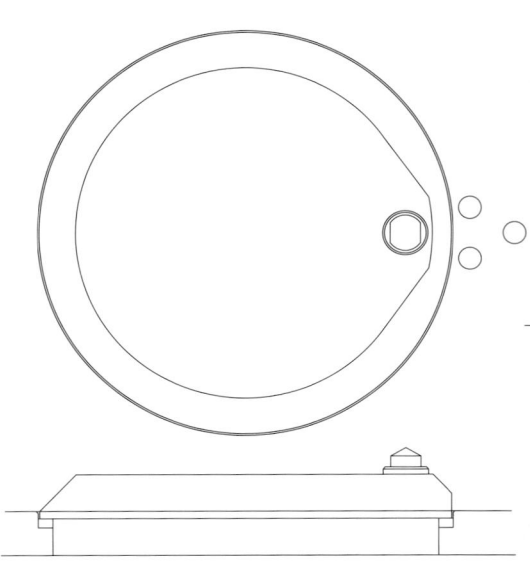

Turmlukendeckel
(Access Hatch)

**Copyright Panzer Tracts 2008**

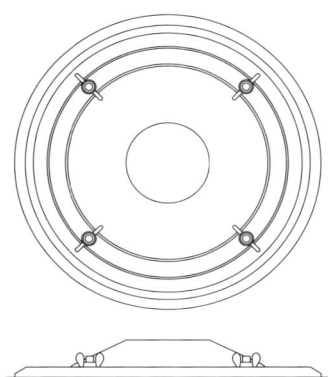
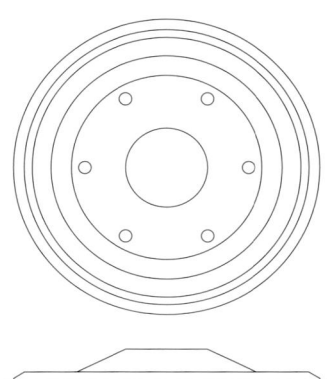
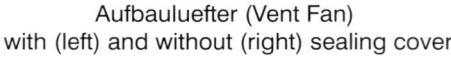

Aufbauluefter (Vent Fan)
with (left) and without (right) sealing cover

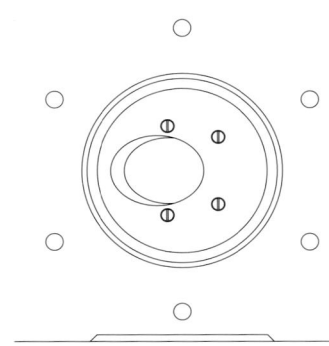

Nahverteidigungswaffe
(Close Defense Weapon)
Dwg.Nr. 021 B 4401.034

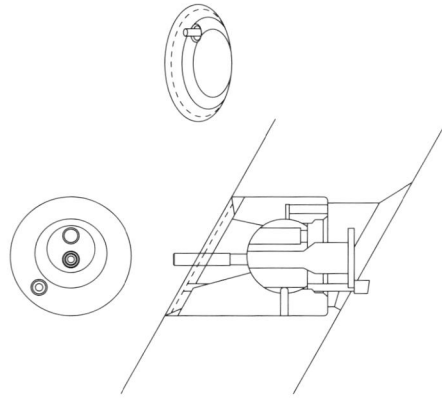

MP Kugelblende
(Machine-Pistol Ball Mount)
with pivoting sealing cap (dashed line)

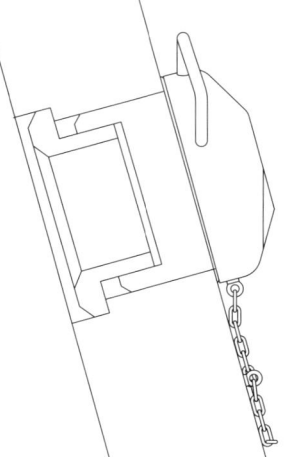
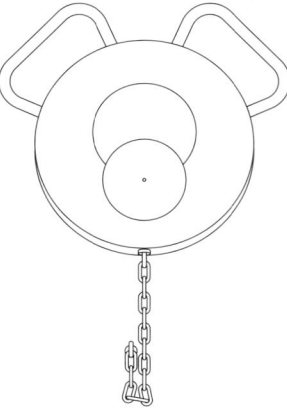

Munitionsluke (Spent Cartridge Ejection Port)
with MP Stopfen (Machine-Pistol Port)

**1/10 Scale**

November 1943: *Zeiss presented a proposal for a horizontal **E-Messer 2.3 m**. The eye piece was located between the turret wall and the gunsight. This location can remain if only a segment is cut out of the turret side in which the gunner's head can fit while range finding. Krupp requested a drawing with measurements for the vertical **E-Messer 1 m** with mounting ring.*

On 23 November 1943, Zeiss wrote to Krupp: *The location of the **Horizontal E-Messer 2.3 m Basis** fits the **Maus Turm** when the gun is depressed at its maximum. A **Panzerkopf** (armor head) protecting the ends of the range finder protrudes out of the turret on the right and the left side. The opening in the turret and the mount for installing the **Vertikal E-Messer 1 m** have been determined and measured.*

### FLA-MG  (Antiaircraft Machinegun)

On 5 February 1943, Wa Pruef 6 wrote to Krupp:
*The recent exchange of ideas with the Reichsluftfahrt-Ministerium resulted in the view that mounting vertical upward firing **Flugabwehrwaffen** on Panzers promises to be successful. The aircraft tactic of dropping bombs and strafing Panzers results in a considerable reduction between the interval when the bomb is released or shots fired and the aircraft flies over the Panzer. As a result, Reichtsminister Speer wants this system not only to be used for the **Pz.Kpfw.Maus** but also quickly installed on the **Tiger**, **Panther**, and **neue Sturmgeschuetze**.*

*Wa Pruef 6 is awarding a contract to immediately design a mount for a **M.G.151 (2 cm)** in **Turm Maus** to the left beside the **Kw.K.** as it was originally intended.*

Krupp sent a telegram to Oberstlt. Crohn (Wa Pruef 6) on 9 March 1943: *The conceptual design for mounting the **MG 151/20** is completed. The vertical weapon is mounted left forward on the turret ring in a bolted on rigid base. The barrel protrudes through a large hole in the roof that is sealed against water by an inserted ring with collars. The barrel can be changed without dismounting the weapon by loosening several bolts. An ammunition bin with 250 belted rounds is located forward and the spent cartridge bag to the left. Electrical firing will be operated by the commander, who has the best all-round vision. When the **Kommandantenkuppel** hatch is closed, the commander's view is restricted to +10 degrees. This must be changed to at least 30 degrres by installing a traversable periscope in the cupola lid with a field of view of 0 to +30 degrees.*

The drawing AKF 81291 proposed mounting of the **MG 151/20** in the **Maus** dated 22Mar43 remained unchanged from the above description with the exception of 80 rounds (instead of 250) in the ammunition bin.

Krupp sent a telegram to Oberstlt. Crohn, Wa Pruef 6 on 24 March 1943, informing him that the **MG 151/20** sent to Krupp from Adlershof had been destroyed in a bombing raid and requesting a new one. Another **MG 151/20** was sent to Krupp on 16 April 1943.

After the wooden model of the **Maus** was shown to Hitler et al. on 14 May 1943, Porsche reported that an all-round traversable **Fliegerabwehrkuppel** with a 3.7 cm gun was requested.

At a meeting in Stuttgart on 21 May 1943 attended by Prof.Dr.Porsche and Krupp:

*The question of installing a **Flieger-MG** that can be aimed to the front or rear results in the following: By dropping the vertical rigidly mounted **Waffe 151/20**, (which Krupp had always advised be done), it is possible to install a forward aimed machinegun in the turret as had been included in conceptual designs already at the end of 1942/early 1943. The machinegun can be aimed to the side by traversing the turret and elevated with the main gun. Independent movement of the machinegun is not possible because of keeping the hole in the turret front as small as possible. Machineguns are not to be mounted in the rear and both sides. Instead **MP-Luken** (pistol ports) are planned. A machinegun in the rear would only have a maximum of 5 degrees traverse due to the tight space between the ammunition racks. It would also have to be removed from this location when the **12.8 cm Kw.K.** is fired.*

*A **Fliegerabwehrkuppel** with its own traverse and elevation can't be installed because of lack of space.*

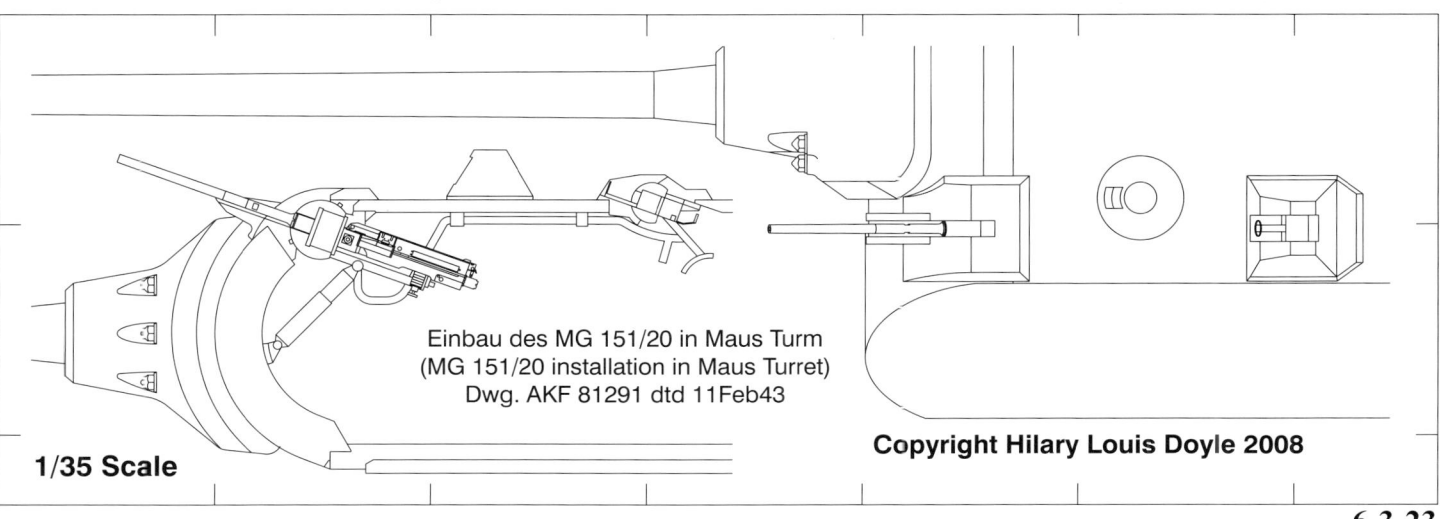

Einbau des MG 151/20 in Maus Turm
(MG 151/20 installation in Maus Turret)
Dwg. AKF 81291 dtd 11Feb43

1/35 Scale

Copyright Hilary Louis Doyle 2008

## MAUS PRODUCTION PLANS

Production of the first five **Maeuschen** was discussed at a meeting on 17 November 1942 attended by Odl. Saur (Min.f.Bew.u.Mun.), General v. Radelmeier and Prof.Dr.Porsche (Panzerkommission), Oberst Thomale and Oberstlt. Bollbrinker (In 6), Oberst v. Wilcke, Oberstlt. Crohn, and Oberbaurat Kniepkamp(Wa Pruef 6), and Obering. Dorn (Krupp): *The highest position (Hitler) has ordered that five **Maeuschen** be completed by May 1943. Porsche declined giving a schedule for the drawings and explained that Krupp would design the **Wanne** (armor hull). Dorn named a deadline of March 1943 for turret and chassis drawings and gave a theoretical minimum time for production of the first **Maeuschen** as an additional 6 or 7 months. As replacements for the **Maeuschen**, three **Porsche-Tiger** are to be converted to **Sturmgeschuetz**. Half a year had been lost because Porsche's negotiations with the bureaucracy have stretched out from June 1942 until today.*

On 19 December 1942, OKH/Wa Pruef 6 awarded Fried.Krupp A.G. Abt.AK, Essen contract SS 006-4574/42 to complete a **Versuchsturm** for the **Pz.Kpfw.Maus** with all the interior components. This was followed on 15 January 1943 by contract SS006-6387/42 awarded by WaPruef6/III to Fried.Krupp AG, Essen for a **Wanne** for a **Versuchs-Fahrgestell Maeuschen** based on directions from Dr.Inf.h.c.Porsche K.G., Stuttgart-Zuffenhausen which was to be delivered to the assembly firm Alkett, Berlin-Borsigwalde. At a meeting with Porsche in Stuttgart on 18 January 1943, it was decided that the **Versuchsfahrzeug Maus** must be completed in September 1943 at the assembly firm Alkett.

At a Panzer-Kommission meeting in Stuttgart on 21 January 1943 attended by Gen.Lt.Ritter von Radlmaier (Panzerkommission) Oberstlt. Holzhaeuer, Oberst von Wilke, Oberstlt. Crohn, Oberbaurat Kniepkamp (Wa Pruef 6), Dir. Freyberg (Alkett), Direktor Dr. Pohl (Skoda), Direktor Dorn (Krupp) Professer Dr. Porsche, Porsche jun., and Chefkonstrukteur Rabe (Porsche KG): *The conceptual design drawings for the **Maus** were shown by Porsche and the planned layout of the vehicle explained as well as plans to acquire the first **Musterfahrzeug** followed by ordering another five. Oberstlt. Holzhaeuer declared that he was ready to increase the contracts already distributed for one **Musterfahrzeug** to six vehicles. Direktor Freyberg from Alkett explained that they were ready to take over production of this vehicle, not only just for assembly but also acquisition or production of all parts. The schedule for completing the first vehicle is September 1943 with another four to be completed by the end of 1943. Hitler wants a production series of 10 per month to follow as quickly as possible.*

A decision to mass-produce 120 **Pz.Kpfw.Maus** was made at a meeting with Porsche, Krupp, Wa Pruef, and R.M.f.B.u.M on 10 February 1943. Krupp was awarded contract SS4911-0210-9801/43 from WaJRue(WuG)VIIId dated 22 February 1943 for 120 armor **Wanne** and 120 **Turm** for the production series. The **Wanne** were scheduled to be delivered to Alkett, Berlin-Spandau at the rate of 2 in November, 4 in December, 6 in January 1944, 8 in February, 10 per month starting in March 1944. Turrets were to be delivered at the same rate, but one month later.

During a meeting at the R.M.f.B.u.M. on 17 February 1943 attended by Hdl. Saur, Panzerkommission (Thomale, Dr. Porsche, Dorn), and representatives from Wa Pruef 6, Alkett, and Krupp (Woelfert): *Hdl. Saur insisted that even though the **Sturmgeschuetz** are still more important than the **Maus**, the **Maus** production schedule must be shortened by 15 days.*

On 23 February 1943. Wa Pruef 6 expanded contract SS4911-0006-6387/42 from one to six **Wanne** for the **Pz.Kpfw. Maeuschen**.

Krupp was also given contracts to produce guns for the **Maus**. Wa J Rue (WuG 2) awarded contract SS 4911-0166-0830/42 H dated 4Mar43 to Fried. Krupp A.G., Essen for 120 **12.8 cm Kw.K. L/55** based on drawing **5-1208** to be delivered at a rate of 1 in August, 3 in September, 6 in October, 5 in November, 5 in December, 7 in January, etc., for installation in turrets being assembled at Krupp.

The first setback to the **Maus** project was caused by a bombing raid on Krupp in Essen. On 11 March 1943, Krupp sent a telegram to Hdl. Saur, Wa Pruef 6, and Porsche: *All of our calculations and drawings for the **Maus** turret and armament were destroyed as a result of the events on 5/6 March. We have already started work on replacement drawings and are attempting in every way to make up for the loss. However, a significant delay in production startup can't be prevented, so that the delivery schedule of the first turret (scheduled for 15 October) will be pushed back at least two months. **Wanne** delivery is on schedule so that there is no delay in completing the vehicles. The completed turrets can be mounted on the chassis in a few days.*

At a meeting in Essen on 16 April 1943, Krupp informed Wa Pruef: *The deadline for completing the **1. Versuchsturm** (first trial turret) must be pushed back from 15 October to 15 November because of a bombing raid. The wooden model has again been burned in the last bombing raid and work on another hasn't started yet because there is no available space or woodworking tools. It will take about 8 weeks to complete.*

The overall production schedule for the **Maus** was reflected in Krupp's plans dated 22 April 1943 to complete operational turrets with guns as follows: **Versuchsturm**: 1 in November 1943, **Null-Serie**: 3 in December and 3 in January, **Leifer-Serie**: first 2 in January 1944, 5 in February, 5 in March, 7 in April, and 10 per month starting in

May 1944.

On 5 May 1943, WaJRue(Wug6)VIIId amended contract SS4911-0210-9801/43 to increase the total number of hulls and turrets for mass production of the **Type 205 "Maus"** from 120 to 135 with the first 2 hulls to be delivered in November, 5 in December, 8 in January, and ten per month starting in February 1944 and the completed turrets to be delivered at the same rate, one month later.

In spite of contracts having already been awarded by the Waffenamt to produce a total of 141 **Pz.Kpfw. Maus**, in June 1943 Generaloberst Guderian (General Inspekteur der Panzertruppen) informed the Panzerkommission that he only intended to allow production of five **Pz.Kpfw. Maus** in order to test their usability and combat value before authorizing series production. On 1 July 1943 the Panzerkommission decided to cut back monthly output of the **Maus** from 10 to 5 per month but to maintain the production series start-up schedule.

On 12 July 1943 Krupp was informed by the Waffenamt that the six **Versuchs** were assigned **Wanne und Turm Nr. 351451 to 351456** and the 135 in the **Sereinausfuehrung** were assigned **Wanne und Turm Nr. 351457 to 351591**.

A bombing raid on Krupp in Essen was successful in killing off the **Maus** series production program. Details of the damage were reported by Krupp on 4 August 1943:

*Wanne Nr.1 is in Mb.10 but further work there is impossible. Alkett has been asked if they can take over working the hull. A **Kulemeyer** vehicle is needed to move it out of Mb.10. Delivery by the end of August can't be met and will be delayed about four weeks.*

*Wanne Nr.2, 3, and 4 are in the Wagenwerkstatt. Work can continue as soon as electricity and compressed air are restored. Wanne Nr.2 can be delivered in about 3 days for working when rail transport is restored.*

*Wanne Nr.5, 6, and 7 can be completed and delivered from the Panzerbau as soon as electrictiy is restored and the crane is repaired. Armor parts have been cut out for **Wanne Nr.8** and 9 and are available in the Panzerbau.*

*Most of the armor plates for 10 **Wanne** have been delivered by the Panzerplattenwalzwerk. Most of the armor plates for another 20 **Wanne** have been rolled. These pieces will be accessible once the rubble has been cleared away.*

*The armor body for **Turm Nr.1** is in 2.m.W. Completion is dependent on getting 2.m.W. back into operation and obtaining/replacing destroyed interior parts. The onset of the bad weather period will hinder production as long as the roof isn't repaired. The schedule for completing the first **Turm** was 15 November, and we now strive for 1 December 1943.*

*The armor parts for **Turm Nr.2, 3**, and 4 are in the Wagenwerkstatt. The armor parts for **Turm Nr.5** to 9 have been flame cut, but mechanical working depends on getting the Panzerbau back into operation.*

On 18 August 1943, Krupp concluded that due to the destruction from the bombing raid, it will take 7 months to restart **Wanne** production and 8 months for **Turm** production. By 25 August 1943, Krupp had been informed by the Waffenamt that only 30 **Wanne** and **Turm** were to be completed predicated on the results of testing the first Maus. **Wanne** delivery was now scheduled as 1

**Above: Maus Versuchs-Fahrgestell Nr.1 being test driven at the Alkett assembly plant. (TTM)**

Sep, 2 Oct, 1 Nov, 1 Dec, 2 Jan, 2 Feb, 3 Mar, 4 Apr, 4 May, 4 Jun, 4 Jul, 2 Aug, 0 Sep and the completed **Turm** scheduled as 1 Nov, 2 Dec, 1 Jan, 1 Feb, 2 Mar, 3 Apr, 4 each May-Sep44.

Krupp was informed by telegram from the Sonderausschuss Panzerfertigung on 27 October 1943: *Hdl. Saur has decided that only one **Maus** is to be completed. All of the manpower, machines, and equipment for **Maus** production are to be immediately employed in increasing other armored vehicle production. The armor is to be used for achieving the ordered increase in **Sturmgeschuetz** production. Please report how much material has been prepared for the **Maus** and which armor plates can be diverted to **Sturmgeschuetz** production. This material must be immediately transferred to Harkort-Eicken.*

As reported by the Panzeroffizier beim Chef Gen. St.d.H. on 4 November 1943, development of the **Maus** was ordered to be canceled in late October 1943. Only one **Maus** was to be completed. On 5 November 1943, the WaJRue canceled contract SS4911-0210-9801/43 for the series production of **Turmpanzer** and **Wanne** for the **Maus**. This was followed by reduction of contract 006-4575/42 from six to one **Turm** on 5 November and reduction of contract SS006-6387/42 from six to two **Versuchs-Fahrgestell "Maeuschen"** on 12 November 1943.

### ACTUAL PRODUCTION AND ASSEMBLY

Krupp reported that the **Wanne Nr.1** was welded together on 7 July 1943. The specified overall width had been exceeded by 17 mm (11 mm on the left and 6 mm on the right). The unfinished **Wanne Nr.1** (still needing machining work) was sent from Krupp, Essen to Alkett on 26 September 1943. Assembly work on the **Fahrgestell Nr.1** was completed at Alkett on 22 December and it was loaded for transport to Boeblingen on 10 January 1944.

Alkett reported on 10 January that assembly work on the **Fahrgestell Nr.2** had begun at Werk Spandau on 8 January 1944. Krupp reported that the **Fahrer-Optik** (periscope) for **Wanne Nr.1** and **2** and the **Graetings** for **Wanne Nr.2** were sent to Alkett on 31 January 1944.

On 7 February 1944, Alkett, Altmaerkisches Kettenwerk G.m.b.H., Werk Spandau, Berlin-Spandau reported: *As a result of higher priority **Sturmgeschuetz** assembly, work on the **2.Maus Fahrgestell** has been delayed and totally halted during the last 14 days. In the meantime the OKH has decided to transfer the entire assembly work to Boeblingen.*

Only partially assembled, the **2.Maus Fahrgestell** (with only the suspension and mechanical brakes installed in an otherwise empty hull) was loaded on 7 March 1944 for transport to Boeblingen, where assembly work was to continue.

**Turm Nr.1** assembled by Krupp in Essen was inspected by Wa Pruef 6 on 16/17 April 1944 and the following items discussed: *Elevation of the **12.8 cm Kw.K.** was satisfactorily measured to be -7 to +24 degrees. Adjustment of the connecting rod for the **T.W.Z.F.1** gunsight must be easier and more accessible. The handles for both hatches must be bent so that the **12.8 cm Kw.K.** can recoil unhindered. The **MP-Kleinstkugelblende** aren't completed*

**This and Opposite Page:** Exterior and interior photos of Turm Nr.1 (the only Maus turret assembled by Krupp) which were included in copy 13 of the secret manual dated 1 July 1944 for the Turm of the Panzerkampfwagen Maus Versuchsgeraet. (KSM)

or installed. An example is being worked on by Krupp. The chain holding the **Munitionsluke** (ammunition port plug) on the turret rear must be stronger. The linkage for firing the **M.G.34** should be relocated so that it doesn't get pinched between the machinegun and the ammunition bin. The vertical **Zeiss Em** (range finder) promised by mid-April hasn't arrived yet; however, the mounting and **Verschlussdeckel** (plug) have already been installed. The periscopes for the **drehbare Winkelspiegel Lagerung** (traversing periscope mount) for the commander haven't been delivered. If more turrets are produced, the **Nahverteidigungswaffe** installed at the rear of the turret should be exchanged with the ventilator in front of it. Although the **Versuchs-Turm** hasn't been accepted, it can be sent to Boeblingen, where any deficiencies can be corrected by Krupp specialists. Twelve **Spgr.Patr.** (two piece) and 12 **Pzgr.Patr.** (two-piece) for the **12.8 cm Kw.K.** are to be sent to Boeblingen for loading trials, where any deficiencies could be corrected by Krupp technicians. This **1.Versuchs-Turm** was sent to Boeblingen on 3 May 1944 and mounted on the **2.Maus Fahrgestell** in June 1944.

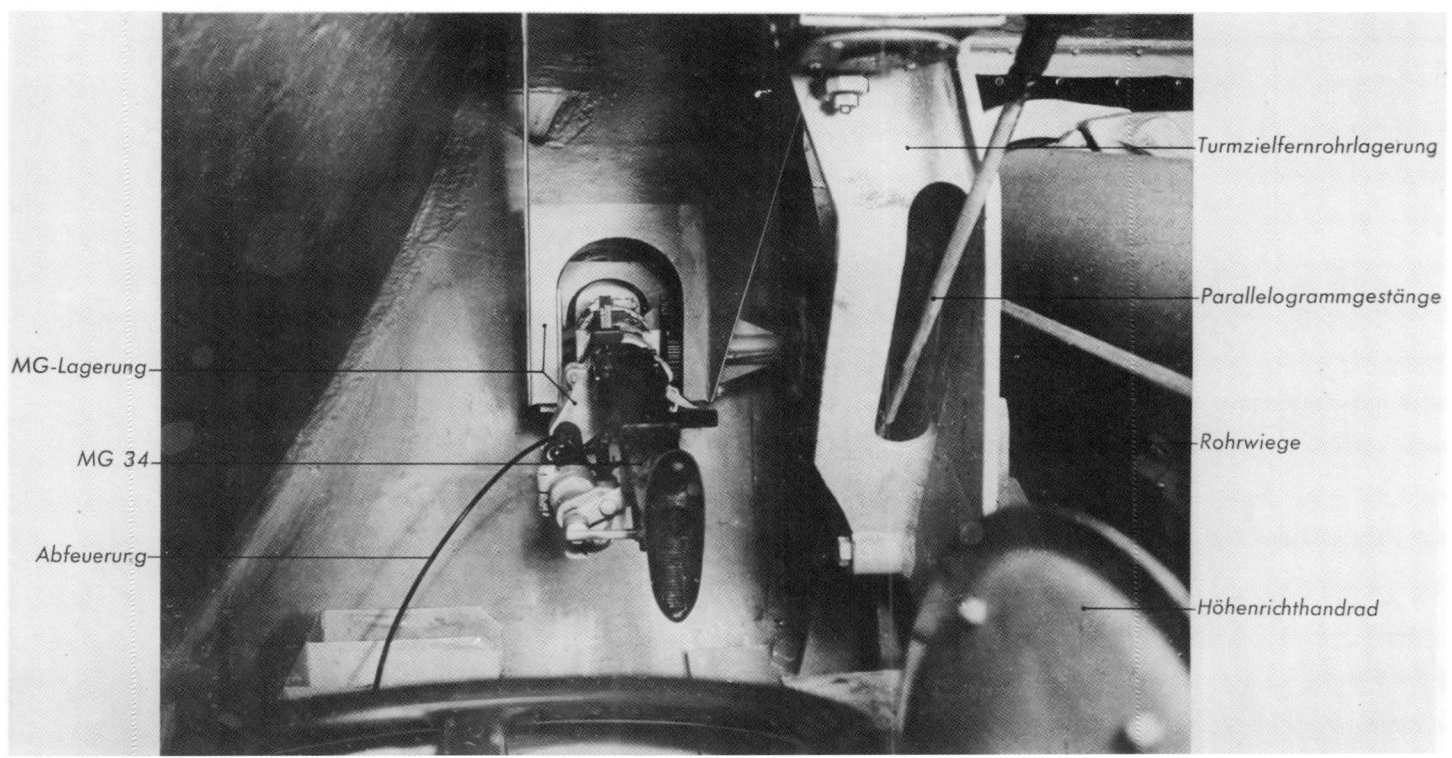

This and Opposite Page: Interior photos of Turm Nr.1 (the only Maus turret assembled by Krupp) which were included in copy 13 of the secret manual dated 1 July 1944 for the Turm of the Panzerkampfwagen Maus Versuchsgeraet. The Drehspiegellagerung was subsequently replaced with a T.Rbl.F.3 (observation periscope). The projectiles for the 12.8 cm Kw.K.44 L/55 were stowed in racks at the rear of the turret separately from the cartridges loaded with propellant. The same single-piece ammunition as used in the 7.5 cm Kw.K. L/24 was stowed in a rack to the right of the 7.5 cm Kw.K.44 L/36. (KSM)

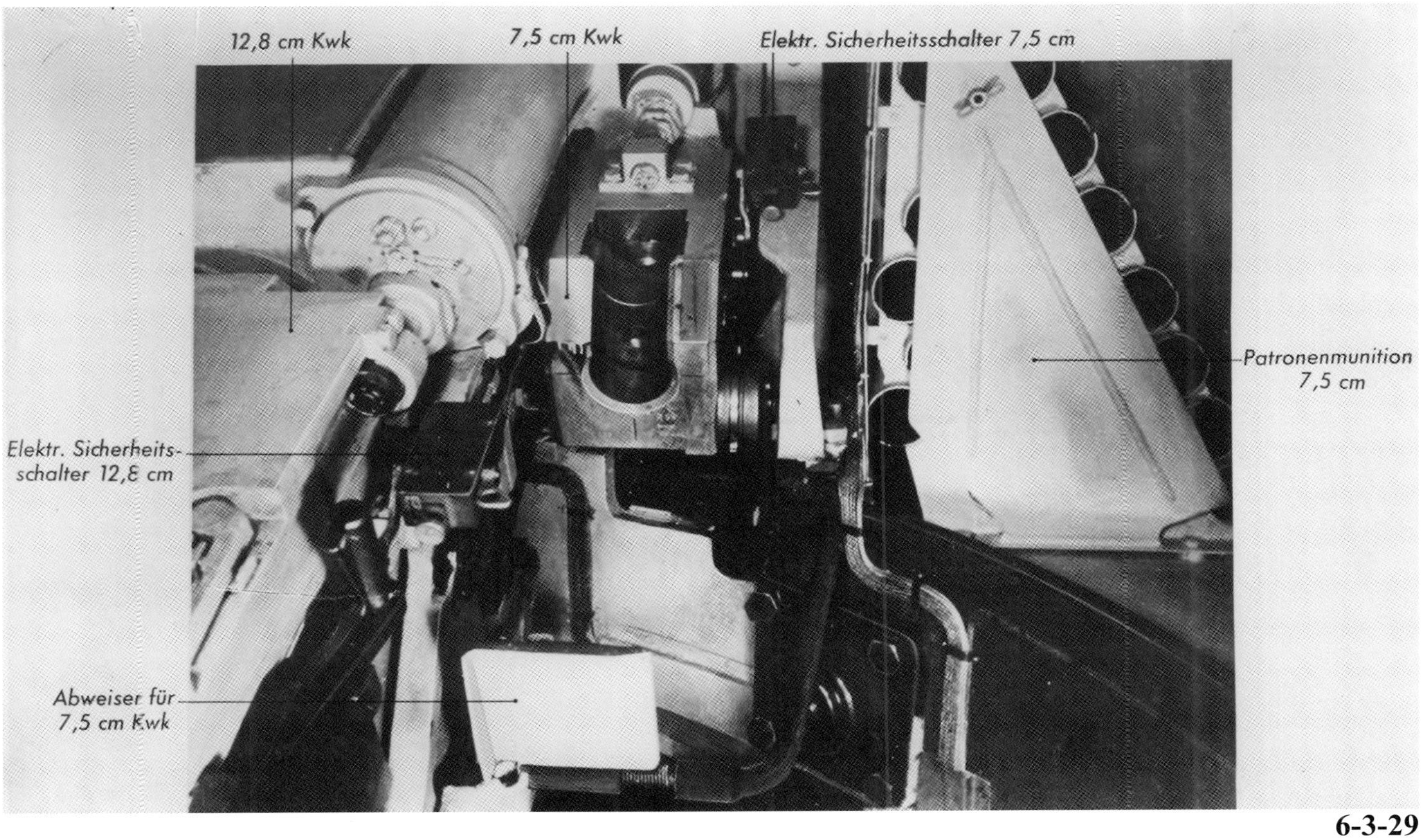

6-3-29

## MAUS TRIALS

On 1 November 1943, Oberst Holzhaeuser, Kniepkamp, and Oberst Crohn from Wa Pruef 6 and Ing.Zadnik from Porsche met to decide upon a **Versuchsprogramm** (trial program) for the **Maus** consisting of: *first factory trials by Porsche, road march driver testing by Kraftfahrversuchsstelle Kummersdorf,* **Tauchversuche** *(submersion trials) and* **Schleppversuche** *(towing trials) by Porsche, and* **Schiessversuche** *(firing trials) in Hillersleben. The order of delivery for testing was the* **1.Maus (Fgst. mit Belastungsgewicht), 2.Maus** *(complete with* **Turm***), submersion equipment, and towing equipment.*

On 9 November 1943, Porsche asked how the **Ersatzgewicht** was to be fastened on the **1.Fahrgestell** for driving trials. The **Ersatzgewicht** was a replacement for the **Maus Turm** that still wasn't finished. Several notches on the bottom of the **Ersatzgewicht** were cut in a circle to center it on the hole for the turret race. It was held by cross pieces that were tightened against the underside of the hull deck.

On 5 February 1944, the first report on the Werkserprobung (factory trials) of **Typ 205/1** (Porsche's designation for **Maus Versuchs-Fahrgestell Nr.1**) in Boeblingen included:

*11-14Jan44 - Transfer of the chassis from Berlin to Boeblingen on a 14-axle* **Spezialtransportwagen** *of the Reichsbahn. Assembly work at Alkett Werk Spandau was completed only so far as needed for the vehicle to be loaded under its own power.*

*14Jan44 - Unloaded using the* **Spezial-Verladerampe***. Drove to the Werkhalle of Panzer-Ersatz-Abteilung 7 Hindenburgkaserne Boeblingen, about 5 km without incident.*

*15Jan44 - Off-road driving and steering trials for about 2 km. As already experienced during steering trials in the tight space in the Werkhalle at Alkett and on the Ruhleben (race track), the vehicle can be steered with great accuracy. This was also demonstrated during the first trials in clay soil where it was easily steered when sinking in over 0.5 meters.*

*16-30Jan44 - Assembly work recommenced to install the* **Fahrschalter** *(driving switch), linkage for foot controls, tachometer and other instruments including electrical and voltage measurement.*

*31Jan44 - Off-road driving trials for about 4.6 km (14 km total trip). As had been awaited, the rubber rings in the roadwheels gave way which had already been experienced during bench testing. Improved replacement roadwheels are already being produced and when completed will be installed in place of the current roadwheels.*

*3Feb44 - The smallest turning circle when driving forward is 14.5 m (measured from the middle of the tracks). It can turn in place with one track going forward and the other in reverse.*

As reported on 26 February 1944, during the Werkserprobung of the **Typ 205/1** in Boeblingen from 4 to 25 February 1944: *During this period the two front and one rear* **Abschleppaugbolzen** *(towing eyes) were welded on as well as various other welding work accomplished to complete the vehicle.*

*7&8Feb44 - Off-road driving with Prof. Dr. Porsche for*

**This and Opposite Page: The 1.Maus Versuchs-Fahrgestell (trial chassis) with the Turm-Ersatzgewicht (turret substitute weight) being test driven at Boeblingen with Prof. Dr. Porsche looking on. (TTM)**

6-3-31

**Above and Below:** The 1.Maus chassis became stuck during a trip through a swampy area because the driver didn't know the area (a region avoided by lighter Panzers). After the mud churned up at the rear was dug out and timbers laid under the tracks, the Maus pulled free under its own power. (TTM)

Left Below and this Page:
As reported on 20 March 1944, the 1.Maus was jacked up to install the improved roadwheels. The drive train components such as the engine, generators, electrical motors, final drive with brakes, and the differential were taken out and dismantled for inspection. (WJS)

Maus Versuchs-Fahrgestell Nr.1 with Turm-Ersatzgewicht

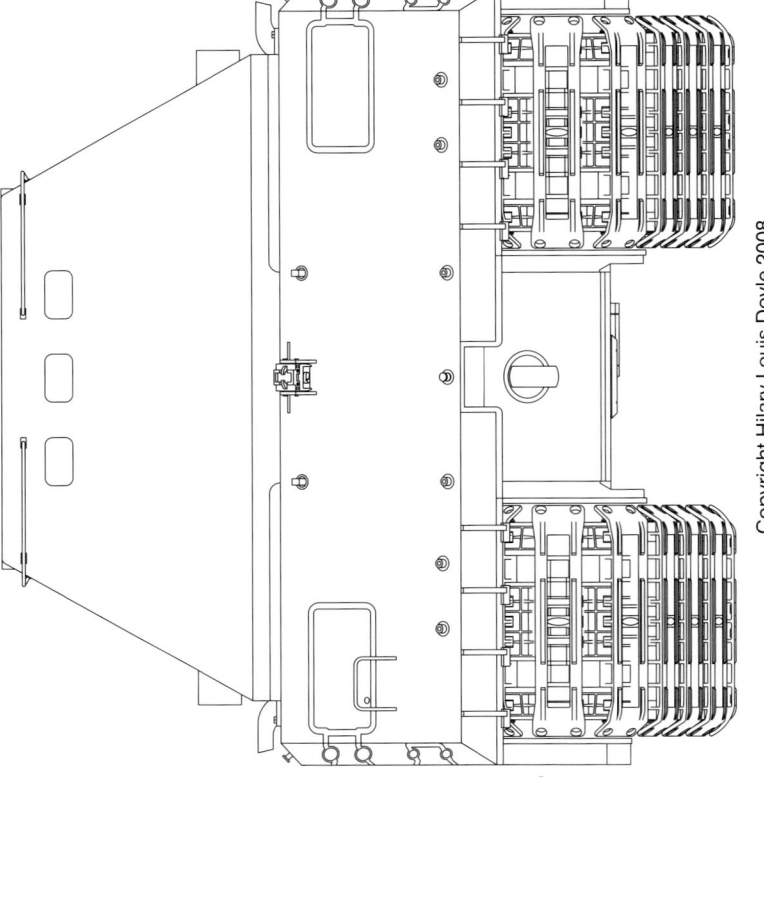

Copyright Hilary Louis Doyle 2008

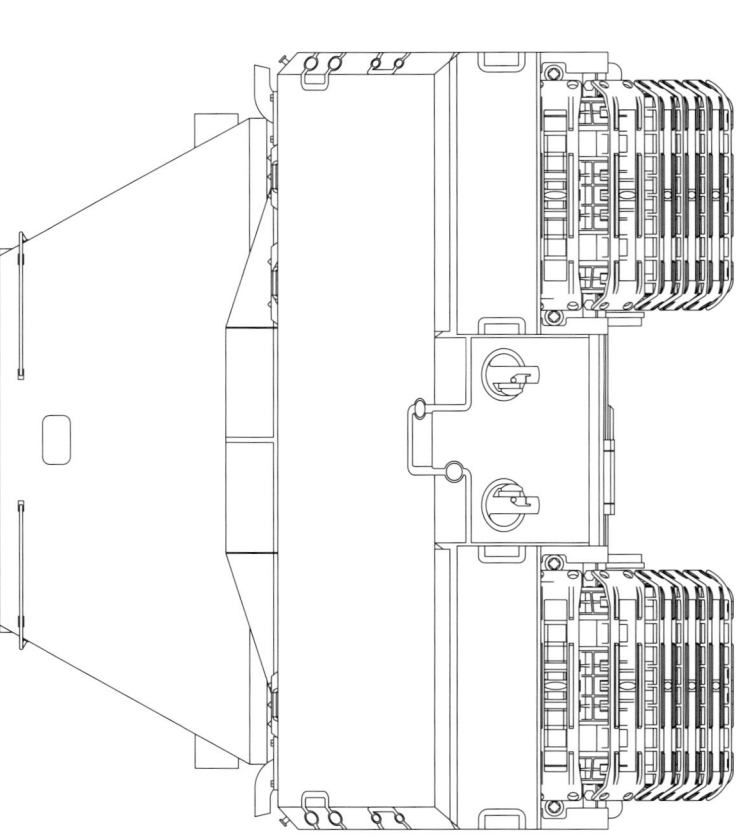

Copyright Hilary Louis Doyle 2008

1/35 Scale

Versuchs-Fahrgestell Nr.1 with Wanne Nr. 351451 has 180 mm thick side armor with an overall width of 3700 mm. The driver's periscope is centered 180 mm to the left of the center line and the rectangle surrounding the driver's hatch and periscopes is recessed 3 mm deep. Rectangular holes that had been cut out of the upper hull rear plate for mounting Flammenwerfer (flame throwers) have been plugged.

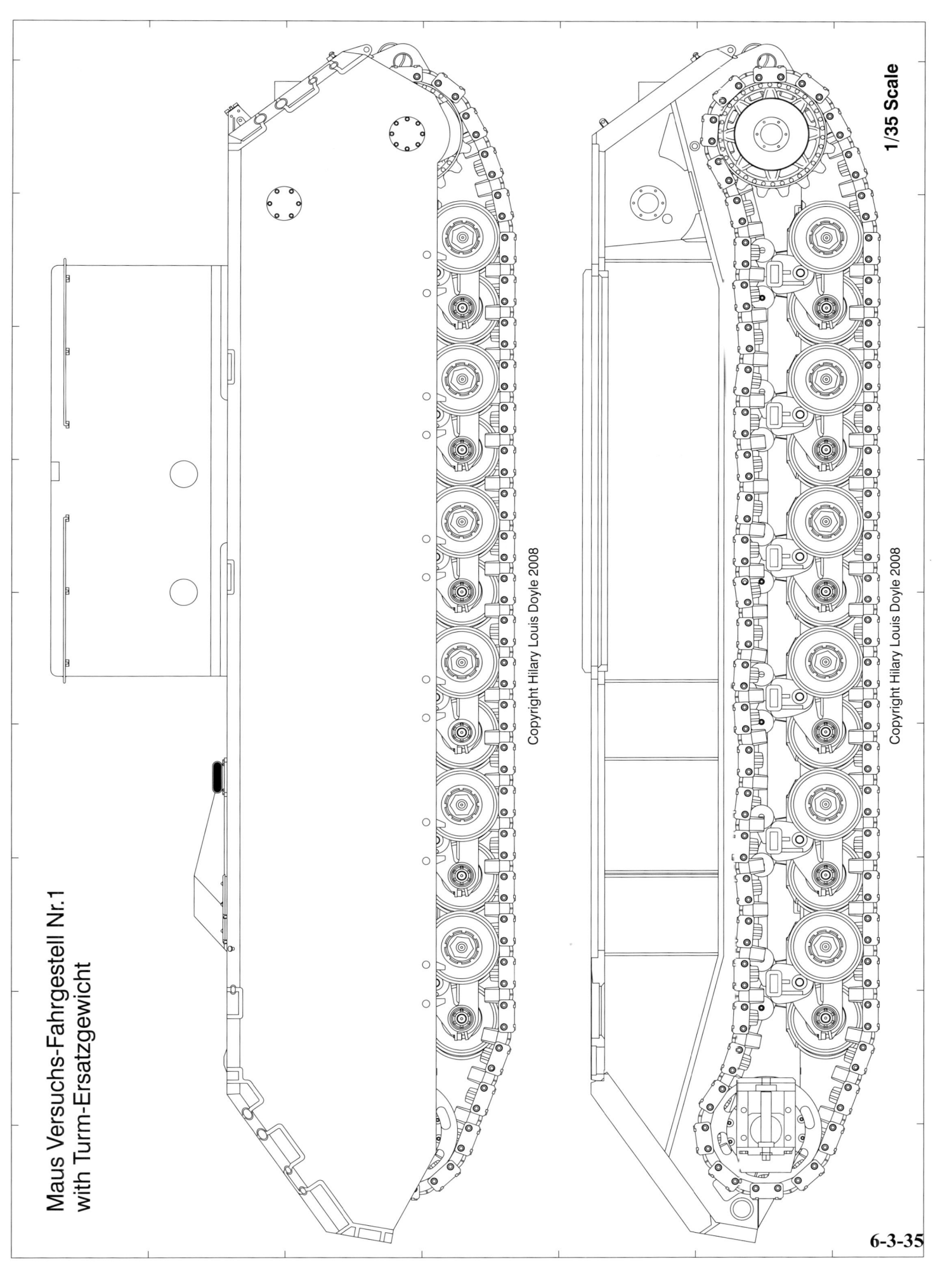

# Maus Versuchs-Fahrgestell Nr.1 with Turm-Ersatzgewicht

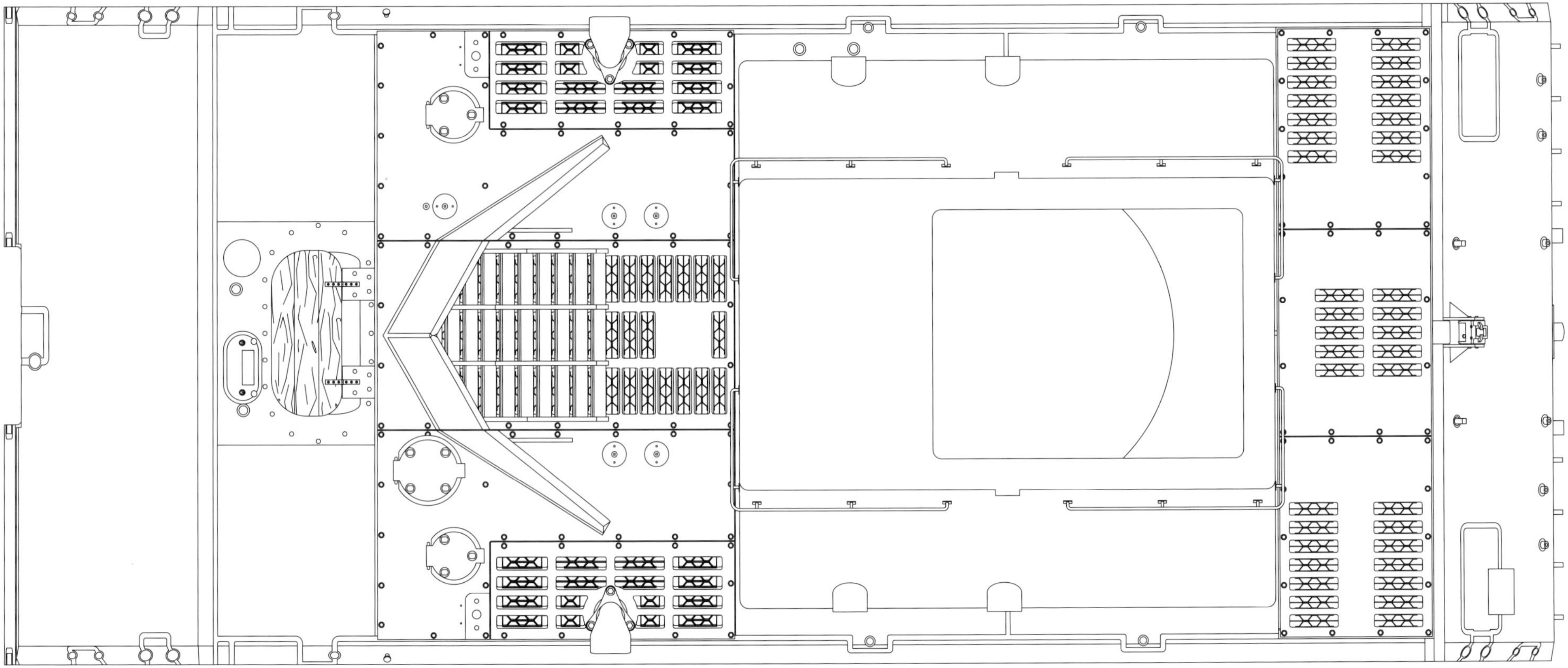

Copyright Hilary Louis Doyle 2008

1/35 Scale

Versuchs-Fahrgestell Nr.1 with Wanne Nr. 351451 has 180 mm thick side armor with an overall width of 3700 mm. The driver's periscope is centered 180 mm to the left of the center line and the rectangle surrounding the driver's hatch and periscopes is recessed 3 mm deep. Rectangular holes that had been cut out of the upper hull rear plate for mounting Flammenwerfer (flame throwers) have been plugged.

## Panzerkampfwagen Maus
### Versuchs-Fahrgestell Nr.2 with Versuchs-Turm Nr.1

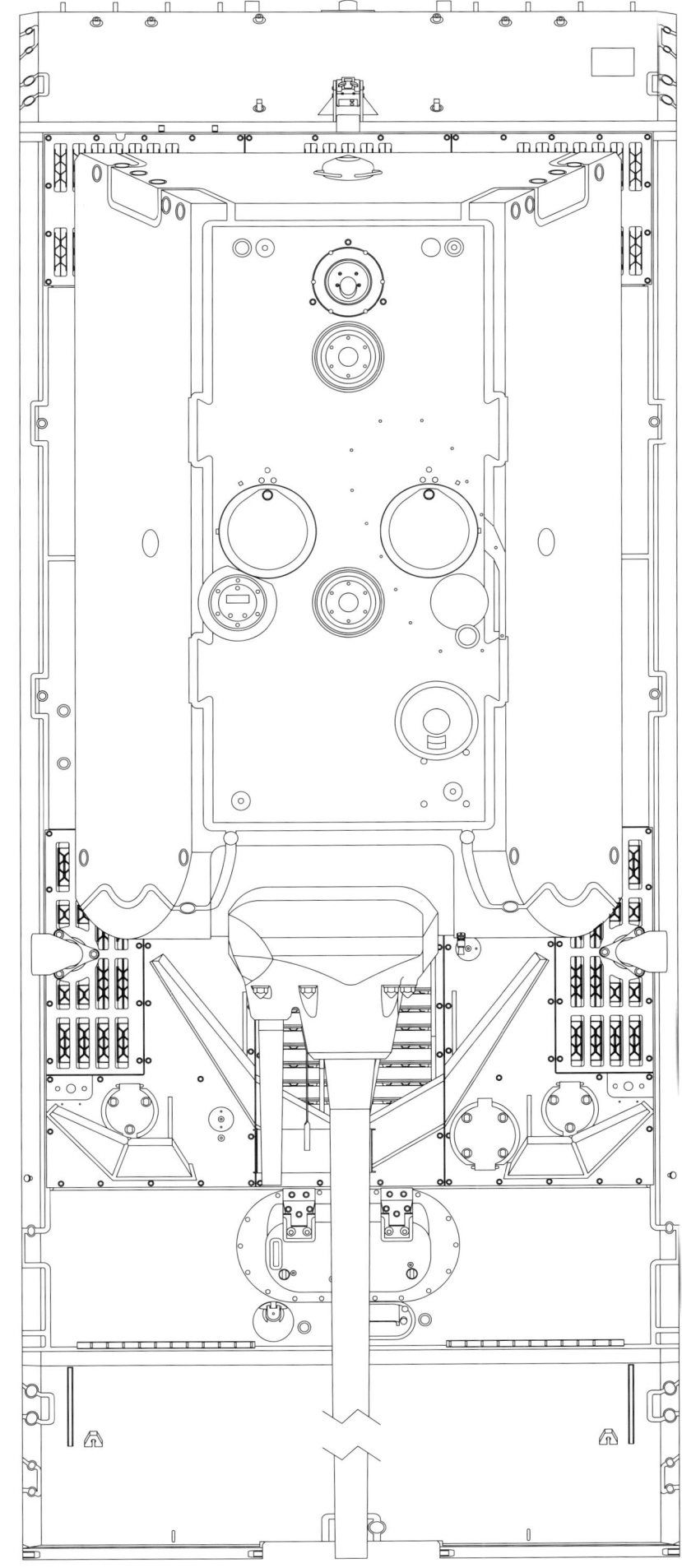

Copyright Hilary Louis Doyle 2008

1/35 Scale

Versuchs-Fahrgestell Nr.2 with Wanne Nr. 351452 has 180 mm thick side armor with an overall width of 3700 mm. The driver's periscope is centered 160 mm to the left of the center line and the oval surrounding the driver's hatch is recessed 3 mm deep. Holes were drilled in the pressure plates on the side of the roadwheels in an attempt to keep the rubber rings from shifting. The Funker Ausblickgeraet (radio operator's periscope) was fitted and additional Schutzleisten (shot deflectors) were added to the left and right on top of the engine compartment deck. The lozenge pattern array of bolt holes on the turret roof was to secure a special Schacht (shaft) over the opened loader's hatch to allow deeper fording.

6-3-37

# Panzerkampfwagen Maus
## Versuchs-Fahrgestell Nr.2 with Versuchs-Turm Nr.1

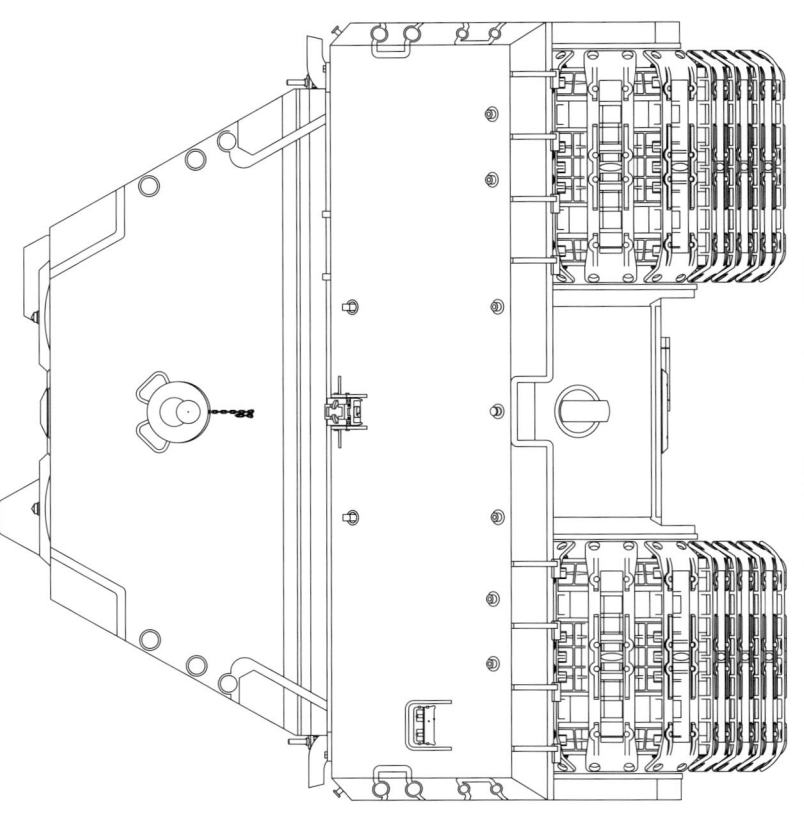

Copyright Hilary Louis Doyle 2008

1/35 Scale

The driver's periscope is centered 160 mm to the left of the center line and the oval surrounding the driver's hatch is recessed 3 mm deep. Holes were drilled in the pressure plates on the side of the roadwheels in an attempt to keep the rubber rings from shifting. The Funker Ausblickgeraet (radio operator's periscope) was fitted and additional Schutzleisten (shot deflectors) were added to the left and right on top of the engine compartment deck.

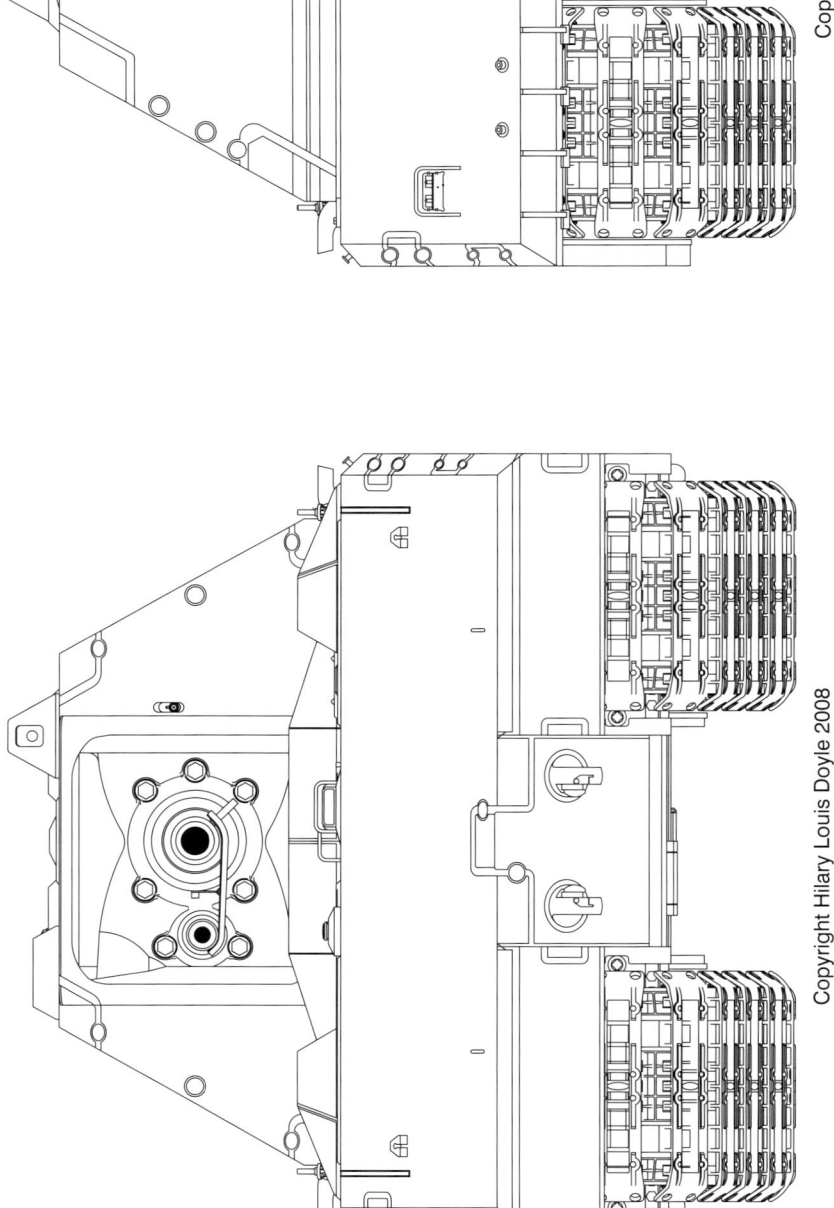

Copyright Hilary Louis Doyle 2008

Versuchs-Fahrgestell Nr.2 with Wanne Nr. 351452 has 180 mm thick side armor with an overall width of 3700 mm.

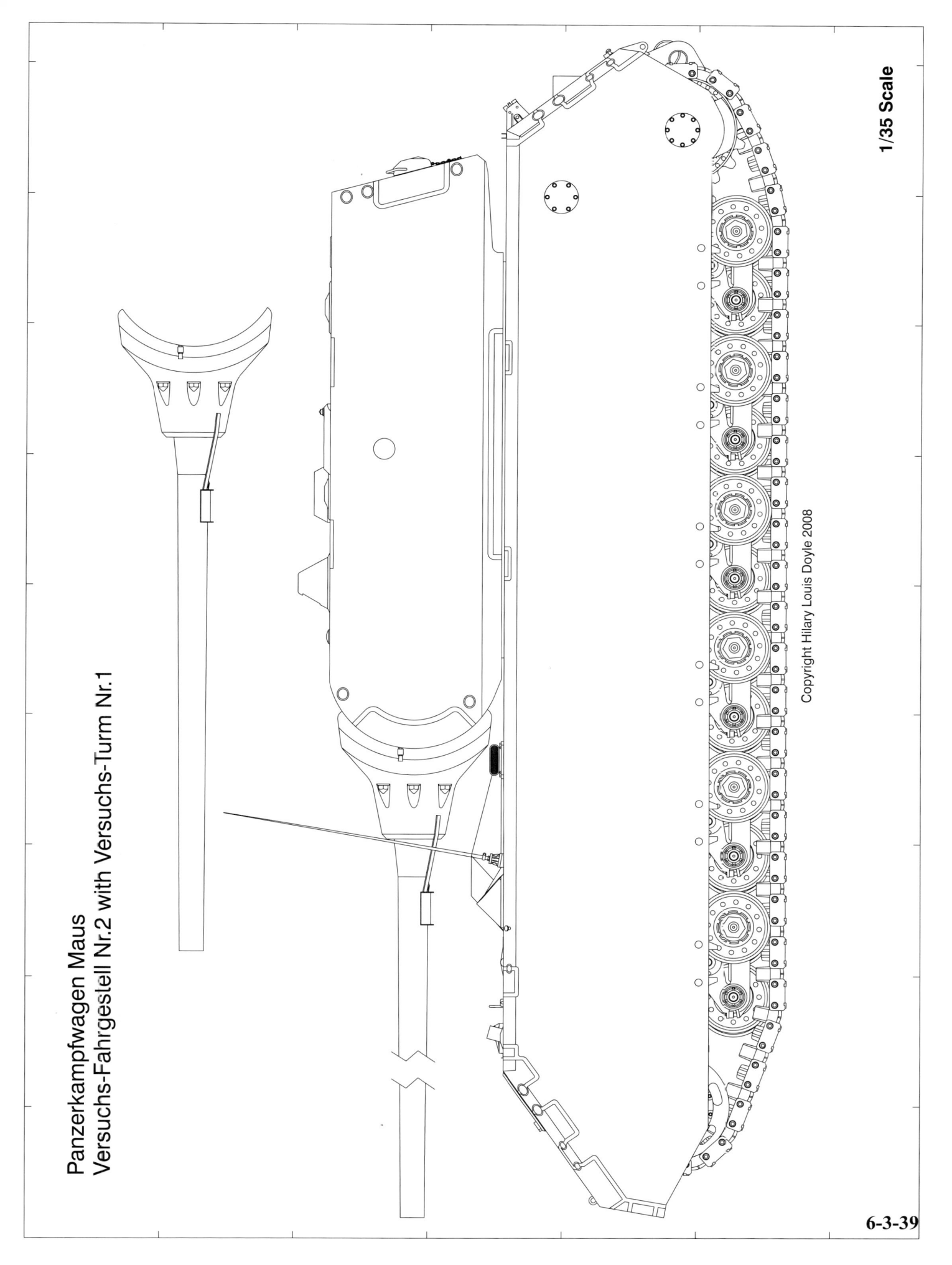

*6.4 km (42.4 km total).*
*8Feb44 - Observed the installation of the Daimler-Benz Motor MB 509.*

As reported on 20 March 1944 for the Werkserprobung in Boeblingen:
*10Mar44 - Unloaded the 2.Maus chassis Typ 205/2. (Porsche's designation for Maus Versuchs-Fahrgestell Nr.2). Fahrzeug 205/2 was only assembled at Alkett Berlin-Spandau as far as components were needed for it to be towable, which were only the complete suspension, tracks, and hand brakes in the hull. Fahrzeug 205/1 towed it to the Werkhalle der Panzer-Ersatz-Abteilung 7 Hindenburgkaserene Boeblingen. The trip with inclines up to 12% on ice covered roads was completed without difficulty.*
*15-17Mar44 - Various attempts made in crossing streams up to 1 m deep with slopes up to 45% were successful. The 1.Maus chassis became stuck during a trip through a swampy area because the driver didn't know the area. As determined later, this region was avoided by the lighter Schulfahrzeuge from Pz.Ers.Abt.7.*

*By digging out the mass of mud churned up at the rear and laying timbers under the tracks, the Maus pulled free under its own power.*

*The 1.Maus was jacked up to install the improved roadwheels. The drive train components such as the engine, generators, electrical motors, final drive with brakes, and the differential were taken out and dismantled for inspection.*

Kraftfahrversuchsstelle Kummersdorf reported on the status of **Maus** trials as of 1 April 1944:

*The available 14-axle Verladewaggon (27 m long rail car) makes rapid loading possible with the side tilted loading area of 6 x 3 m. The associated Verladerampe and Verladestrasse are presently too weakly built and bent during trials.*

*The Motor MB 509 (created from the Flugmotor DB 603) needs fuel with an least 77 octane. For driving trials commercial gasoline is mixed with Flugkraftstoff B 4 (1/3) or C 3 (1/5) or Blei-Tetra-Aethyl (0.09%).*

*Driver's vision is poor. This was temporarily improved by adding cushions, but this makes it impossible to use the driver's periscope and therefore it is untested.*

*Steering ability on firm ground and slippery clay is good due to the separate drive for each track and the curved ends of the track links. Mechanical brakes are sufficient to halt the vehicle.*

*Necessary modifications include: new roadwheels are being installed because the rubber rings fail in the original roadwheels, the spindle in the track tensioner jams because of track hits and plans are to redesign it, too much power is needed to shift into road gear using the hydraulic shifter and therefore it will be redesigned for hand shifting, and it is necessary to make it easier to install the transmission, engine, and generators.*

The **Versuchs-Turm Maus I** arrived at Boeblingen on 3 May and was unloaded on 4 May. Covered with a tarp in the open, it was guarded at the Boeblingen airfield. The rest of the electrical equipment still needed to be installed before it was mounted on the chassis.

As reported in the progress report from Boeblingen for 7-16 June 1944: *The turret was set onto the chassis during the night of 7/8 June for the planned inspection by Generaloberst Guederian; however, he did not show up. The gap between the turret ring and the vehicle was sealed with Terosenkitt. The gap between the bottom edge of the turret and chassis deck was 17 to 22 mm on the right side with the turret lowered and 23 to 28 mm on the left side with the turret raised.*

**Above: Maus Versuchs-Fahrgestell Nr.2 was towed off the special 14-axle Verladewaggon (27 m long rail car) by the Maus Versuchs-Fahrgestell Nr.1 at Boeblingen on 10 March 1944. (Porsche)**

As reported on 27 June 1944: *The **Drehspiegel-einsatz** (periscope) for the commander arrived in Boeblingen. After installation it was determined that it had been made for installation of the periscope at 90 degrees instead of 60 degrees and therefore it was impossible to observe the surrounding terrain. Krupp requested that a **T.Rbl.F.3** for the **Versuchs-Turm** be sent to Boeblingen in order to complete the mounting. **T.Rbl.F.3 Nr.54404** was to be sent to Porsche in Boeblingen on 24 July.*

During the period from 23 June to 2 July 1944: *The **Maus** was being repaired so that it wasn't possible to test the traverse mechanism off road. The elevation mechanism had a play of 0.25 degrees, and the spindle has a dead zone of 6 degrees. An additional 26 rounds of 7.5 cm ammunition were stowed in the turret (bringing the total up to 85). The right half of the turret platform is removable for easier access to the ammunition stowed in the hull.*

During the period from 10 to 17 July 1944: *The traverse mechanism was tested on a 10 degree slope. After the slip clutch was tightened by about another 2000 kg torque, it took about 30 kg force on the handwheel or auxiliary crank to traverse the turret. Traverse by the electrical drive was not possible because the Siemens **Umformer** (motor/generator set) wasn't functioning.*

*The two **MP Kugelblenden** that just arrived on 12 July 1944 were not usable because they were deformed during test firing. Two **MP-Kugelblende** for the **Versuchs-turm Maus** were already available, but the ball is cracked in one of them.*

As reported for the period from 12 to 15 July 1944: *After the Siemens **Umformer** was remounted on rubber, the electrical traverse intermittently failed. The **Umformer** again made an unusual noise in spite of being cushioned by rubber.*

*Driving trials succeeded without incident. The ground was in good condition, so there wasn't any unique stress on the suspension. The **Maus** turns with difficulty. The high ground pressure results in cobblestones being torn out in curves that are driven several times. Fuel consumption is about 350 liters per 10 km.*

*The gasoline engine and tracks have special problems. Apparently there is valve damage in the engine and it should be taken out again. The **Plattenkette** (plate track)*

**Right:** The only Versuchs-Turm assembled by Krupp was mounted on the Maus Versuchs-Fahrgestell Nr.2 at Boeblingen. This second chassis was different from the 1.Versuchs-Fahrgestell in numerous details, including two additional armor deflectors on the deck, headlights, the convoy tail light, and the Daimler-Benz engine. (WJS)

**Above and Below:** The 2.Versuchs-Fahrgestell with the 1.Versuchs-Turm being test driven at Boeblingen. Holes were drilled in the pressure discs on the roadwheels to prevent the rubber rings from being pushed out of position under the tremendous weight load. (WJS)

Above: After being sprayed with camouflage paint (base coat of Dunkelgelb RAL 7028 with Olivgruen RAL 6003 and Rotbraun RAL 8017 stripes), the complete Maus Versuchs-Fahrzeug (Fahrgestell Nr.2 and Turm Nr.1) still had roadwheels with holes drilled in the pressure dics. (WJS)
Below: These roadwheels had been replaced before this Wa Pruef 6 photo was taken. (WJS)

*first installed has proven to be unusable and replaced with a new **griffigere Gleiskette** (track with cleats) from Skoda. Several links in the new track broke under stress. It takes 8 hours for six men to replace the tracks in the Werkstatt.*

On 19 August 1944, Krupp informed Porsche that the Waffenamt had ordered work on the **Typ 205** to be halted and all of the Krupp workers were being called back for other higher priority work.

On 1 December 1944, Daimler-Benz responded to an inquiry about the **MB 517 Motor**: *The **MB 517 Motor** ordered by the OKH is still here in our factory and can be completed at the earliest in 2 weeks. This engine shouldn't be given away because the model **MB 507** is no longer being produced and there aren't any replacement parts available. Another **MB 517 Motor** installed in the **Porsche-Panzer** is still in Boeblingen and is needed there for further trials.*

Both the **Maus** with turret and the **1.Versuchs-Fahrgestell** with **Turm-Ersatzgewicht** were transferred to Kummersdorf for testing in late 1944. Orders for the activation of the unit at Kummersdorf do not list a **Maus** among the operational Panzers. The **Versuchs-Maus** with turret was blown up at the end of the war. The Russians recovered **Versuchs-Turm Nr.1**, mounted it on the **1.Versuchs-Fahrgestell** and sent this "amalgamated" **Maus** to Kubinka.

**Above: The Maus Versuchs-Fahrgestell Nr.1 with Turm-Ersatzgewicht (substitute turret weight) in front of the Maus Versuchs-Fahrgestell Nr.2 with Versuchs-Turm Nr.1. (WJS)**

## RESTART MAUS PRODUCTION

On 13 March 1944, Prof. Dr. Mueller (Krupp) reported that it is possible that restarting **Maus** production will come up for discussion and asked for documentation on how many **Wanne** and **Turm** plus armor material are available and a proposed production schedule. On 18 March 1944, Krupp, Essen reported that seven **Maus-Wanne** have been worked by the Panzeraufbau and sufficient armor plates are available for another eight **Wanne**. Armor material for about 30 **Wanne** and **Turm** bodies had been rolled and cut/worked for about 15 **Wanne** and 9 **Turm**. A quick restart of production at Krupp, Essen facilities wasn't possible because they were fully loaded with other work.

On 23 March 1944, Oberst Holzhaeuer (Wa Pruef 6) informed the Gen.Insp.d.Pz.Tr.: *As related by Prof. Porsche, Hitler has ordered accelerated driving trials and to resume development of the **Maus**. In addition, Porsche has contacted Krupp for delivery of a second **Maus I Turm** and the first **Maus II Turm**. We request orders to clarify if the decision to complete only two **Fahrgestell** (one with a **Turm**) has been rescinded.*

In response to a question on 1 April 1944, whether delivery of armor hulls and turrets can restart to meet the following schedule: two **Wanne** per month starting in July and one **Turm** in June followed by two per month starting in July, Herr Talman (2.mech.Werkstatt at Krupp, Essen) replied: *Production starting with **Maus Nr.8** at a rate of*

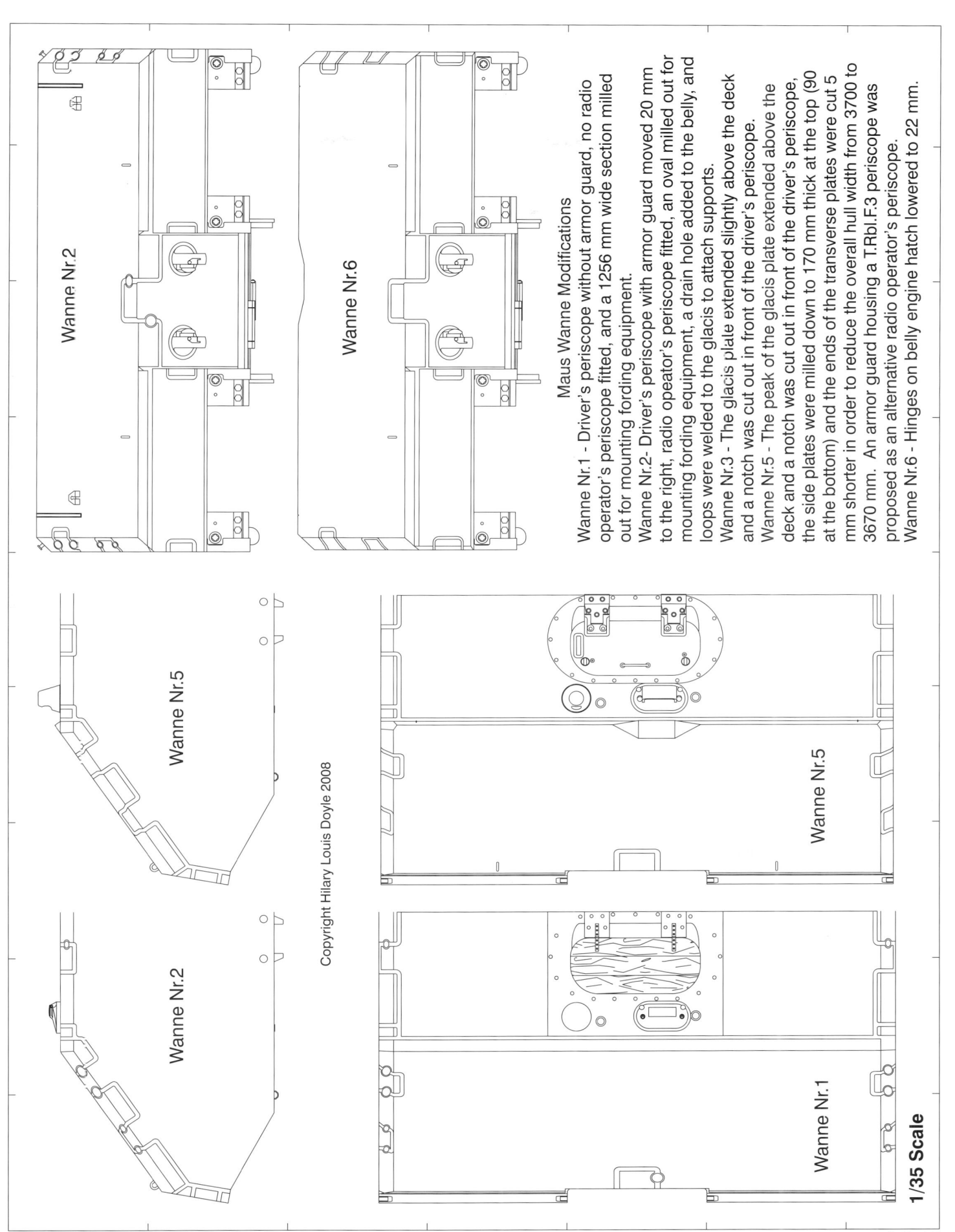

## Maus Wanne Modifications

Wanne Nr.1 - Driver's periscope without armor guard, no radio operator's periscope fitted, and a 1256 mm wide section milled out for mounting fording equipment.
Wanne Nr.2- Driver's periscope with armor guard moved 20 mm to the right, radio opeator's periscope fitted, an oval milled out for mounting fording equipment, a drain hole added to the belly, and loops were welded to the glacis to attach supports.
Wanne Nr.3 - The glacis plate extended slightly above the deck and a notch was cut out in front of the driver's periscope.
Wanne Nr.5 - The peak of the glacis plate extended above the deck and a notch was cut out in front of the driver's periscope, the side plates were milled down to 170 mm thick at the top (90 at the bottom) and the ends of the transverse plates were cut 5 mm shorter in order to reduce the overall hull width from 3700 to 3670 mm. An armor guard housing a T.Rbl.F.3 periscope was proposed as an alternative radio operator's periscope.
Wanne Nr.6 - Hinges on belly engine hatch lowered to 22 mm.

1/35 Scale

The second turret armor body (Turm Nr. 351452) (Above) and the third armor hull (Wanne Nr. 351453) (Below) photographed at Krupp, Essen after the war. This backs up Krupp's reports that only two hulls had been delivered to Alkett for assembly and only one turret had been assembled by Krupp. Another four armor hulls (Wanne Nr. 4 - 7) and five turrets (Turm Nr. 3 - 7) had been welded together at Krupp, Essen before the planned mass production of 135 Maus was abandoned due to the destructive effects of an Allied bombing raid on Krupp, Essen before 4 August 1943. (TTM)

# Panzerkampfwagen "Maus"
## Typ 205/2, Versuchs-Fgst.Nr. 2

**Weapons Data:**
In Turret: 1 - 12.8 cm Kw.K.44 (L/55)
1 - 7.5 cm Kw.K.44 (L/36)
1 - 7.92 mm M.G.34
Elevation: -7, + 23°
Traverse: 360° (electric and hand)
Gun Sight: T.W.Z.F.1 (3x, 10°)
graduated to 4000 m for Pzgr.

**Ammunition:** 68 - 12.8 cm, 100 - 7.5 cm
1000 - 7.92 mm

**Crew:** Pz.-Fuehrer (commander)
Richtschuetze (gunner)
2 Ladeschuetzen (loaders)
Fahrer (driver)
Funker (radio operator)

**Communication:** Fu 5 and intercom

**Measurements:**
Length, overall: 10.085 m
Length, w/o gun: 9.034 m
Width, overall: 3.700 m
Height, overall: 3.649 m
Firing Height: 2.774 m
Wheel Base: 2.330 m
Track Contact: 5.880 m
Combat Loaded: 188 metric ton
Fuel Capacity: 1600 & 1000 liters

**Automotive Capabilities:**
Maximum Speed: 20 km/hr
Avg. Road Speed: 18 km/hr
Cross Country: ?? km/hr
Range on Road: 160 km
Cross Country: 62 km
Grade: 35°
Trench Crossing: 3.5 m
Step: 75 cm
Fording Depth: 200 cm
Ground Clearance: 57 cm
Ground Pressure: 1.45 kg/cm$^2$
Power Ratio: 6.4 HF/ton
Steering Ratio: 2.52
Turning Circle: In own length

**Automotive Components:**
Motor: Daimler-Benz MB 517
V-12, water-cooled
44.5 liter gasoline
1200 HP @ 2500 rpm
Transmission: 2 electric generators
driving two electric motors
Steering: Electric control
Drive: Rear sprocket
Roadwheels: 12 x 2 per side
Tires: 550 mm dia. Steel
Suspension: Volute springs
Track: 1100 mm wide dry pin
Links per side: 56 & 56

## Armor Specifications for the Maus

Armor thickness in mm/angle from vertical

Tolerances for plate thicknesses -0 to +5%

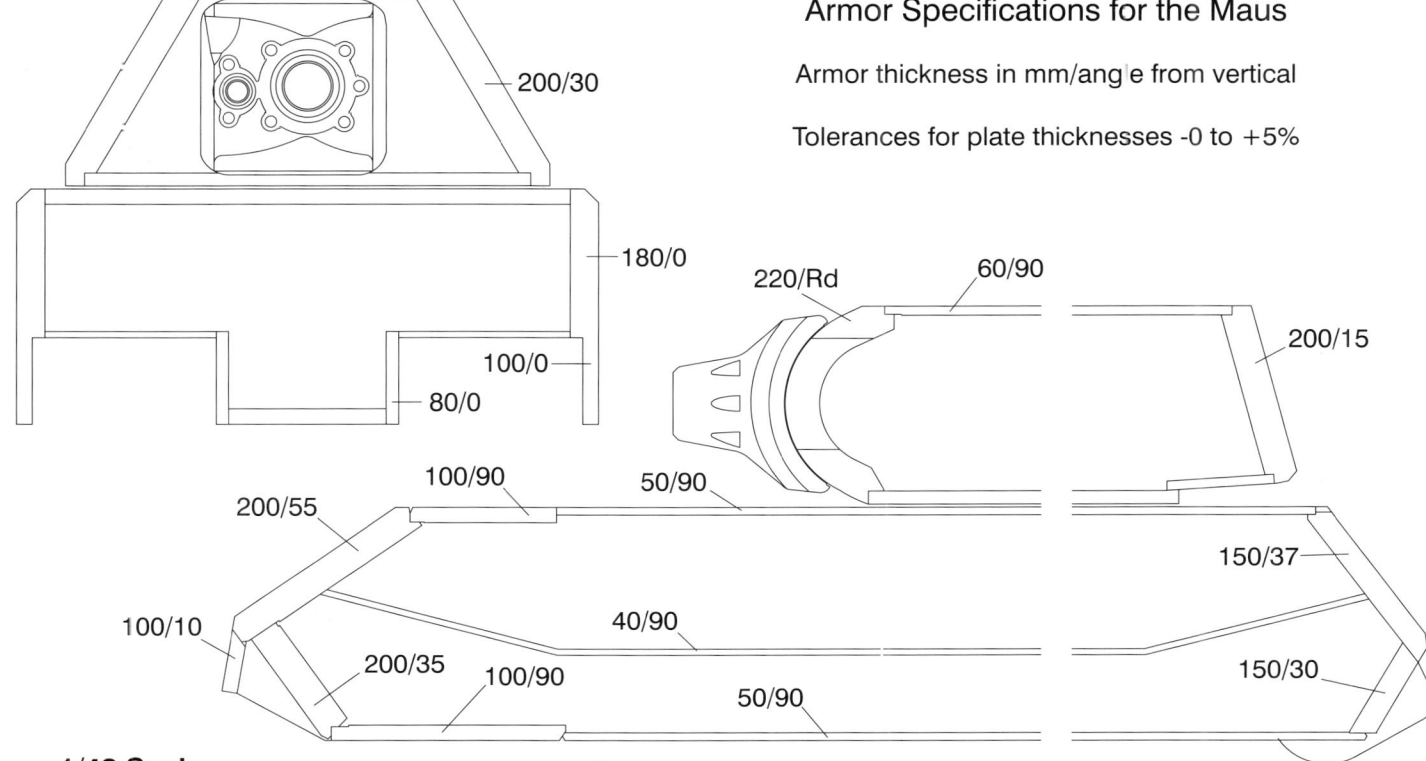

1/48 Scale

*one or two per month is possible only if Panzerbau 1 and Wawe obtain an additional 200 workers. Panzerbau 2 can work two **Wanne** per month but needs a quick decision before the capacity is utilized for another program. The 2. mech.Werkstatt can produce two **Maus** turrets by completing only three instead of five **17 cm Kanone**. If all five **17 cm Kanone** are to be completed as well, they need 27 additional specialists.*

*Delivery starting with **Maus Nr.8** can begin about seven months after a decision is made because new material must be requisitioned. Output of **Wanne Nr.3** to 7 and **Turm Nr.2** to 7 can be completed much earlier because most of the material is available and the armor bodies have been welded together already.*

On 25 July 1944, Krupp asked Wa Pruef for a decision on what to do with **Wanne Nr.3** to **6** left over from contract SS 006-6387/42, because hindrance to ongoing production couldn't be tolerated. Wa Pruef 6 responded on 27 July that the four **Wanne** for the **Versuchsfahrgestell "Maus"** can be scrapped.

## MAUS II

In March 1944, Porsche attempted to restart the **Maus** program and Krupp was involved in creating an improved turret, known as the **Maus II Turm**.

On 8 April 1944, Wa Pruef 6/Pz.II awarded Krupp contract SS 4911/0006/3040/43 to produce a 1:5 scale wooden model of a **Maus II Turm** with the 7.5 cm mounted above the 12.8 cm gun. Wa Pruef 6 met with Krupp on 16/17 April to discuss improvements including a larger turret ring, ventilation openings in the deck of the chassis, installation of a horizontal range finder that was 1.9 to 2 meters long instead of 2.1 meters, a fume extractor for the 12.8 cm ammunition bin, and a redesigned **7.5 cm Kw.K**. with a horizontal breech for the **Maus II**. The full-scale wooden model of the **Maus** turret was to be reworked with **Maus II** features.

Krupp sent drawing Bz 3269 of a **Maus II Turm** with a sloped (instead of rounded) front to Porsche on 15 March 1944 for Porsche to incorporate conforming changes into the hull design.

On 15 May 1944, Wa Pruef 6 modified contract SS 4911/0006/3040/43 to add a 1:10 scale wood model of the **Maus II Turm** including a horizontal range finder and a pivoting gun in accordance with Krupp drawing Bz 3250.

Krupp informed Wa Pruef 6 that they had just started work on the scale models for the **Maus II Turm** on 12 August 1944, and were using components from the full-scale model for the **Maus I Turm** which was located in Kummersdorf.

## 15 cm und 17 cm Sturmgeschuetz on both Maus and E 100 Fahrgestell

On 9 May 1944, during a meeting at Krupp, Obering. Schmidt (Porsche) was presented a set of overview drawings of a **15 cm L/63 und 17 cm L/53 Sturmpanzers** created by Krupp that was to be developed by Porsche on their **Maus** in competition with the **E 100** from Adlerwerke.

During a discussion at Porsche on 17 May 1944 on the subject of a new **Sturmgeschuetzaufbau** (superstructure): *Obering. Hendel related that Krupp will do anything to ensure that the **15 cm L/63** gun is used instead of the **17 cm L/53**. The Porsche design to add a **Flakaufbau** was recently proposed to the Waffenamt by Krupp. This idea was turned down because this **Sturmgeschuetz** is to be escorted by **Spezialwagen**.*

Obering. Hendel made the following notes about the **15/17 cm Sturmgeschuetz auf Mausfahrzeug** on 17 May 1944: *The Porsche design is just in the conceptual stage. Because of the higher **Maus** chassis (in comparison with the **E 100**), the superstructure roof exceeds the rail loading profile. Krupp pointed out the observation instruments mounted in the superstructure and is to send a completed overview drawing for Porsche to determine if a **3 cm Flak-Turm** can be mounted. Because the **Flakturm** interferes with the main gun recoil, Porsche requested information about the circumstances in which the recoil could be shortened. A total of 85 rounds of ammunition were to be carried in the turret.*

On 28 May 1944, Krupp was asked to create a 1:5 scale wooden model of a **15 cm** or **17 cm Kanone auf E 100 Fahrgestell** as a **Studienobjekte** to clarify the space, crew, and ammunition stowage questions.

On 21 July 1944, Oberst Crohn (Wa Pruef 6) wrote to Krupp in regard to the **Sturmgeschuetz 15 cm L/68**: *Reichsminister Speer sent a letter dated 10 July with information that due to the current situation Hitler has ordered a halt to development of all armored vehicles with heavy guns. As arranged by Oberst Holzhaeuer, the wooden model of the **Sturmgeschuetz 15 cm** built by Krupp is to be shown to Generaloberst Guderian. Further development of this **Sturmgeschuetz** by Wa Pruef 6 has ceased.*

## MAUS TURMSTELLUNG

During Hitler's conference with Speer on 30Sep/1Oct43, as Item 14: *Because the current capacity for casting steel isn't sufficient to allow production of heavy **Panzertuerme** with 12.8 and 15 cm guns for fortress lines, investigate if the series production **Maus Turm** (that can be built with an option of either the 12.8 or 15 cm gun, but whose roof must be strengthened) can be used as a conditional replacement and in what number these turrets can be delivered after a certain start-up period.*

Krupp prepared drawing Bz 3186 dated 2 November 1943 of a **Turm "Maus" fuer ortsfesten Einsatz** (**Maus** turret for a fixed installation).

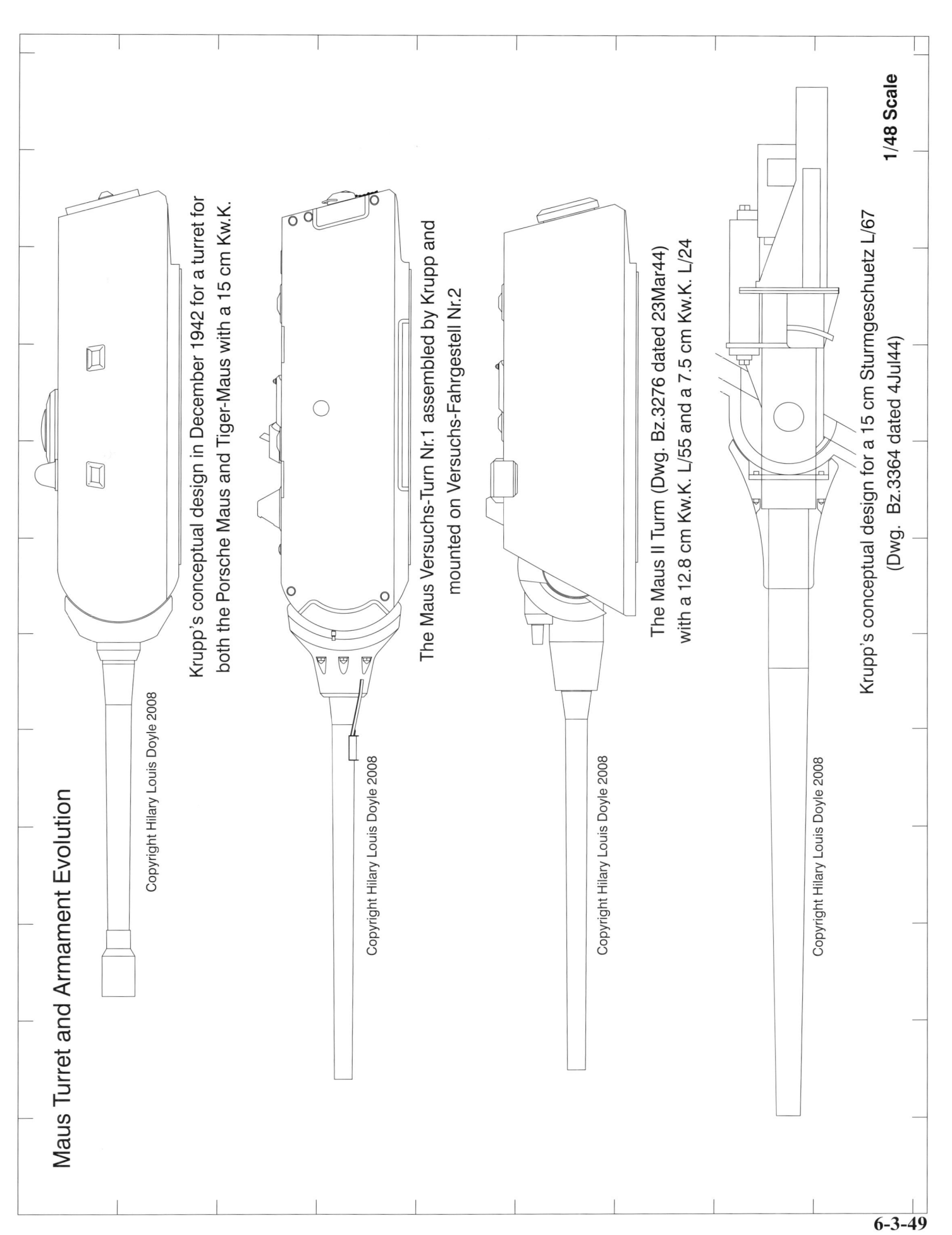

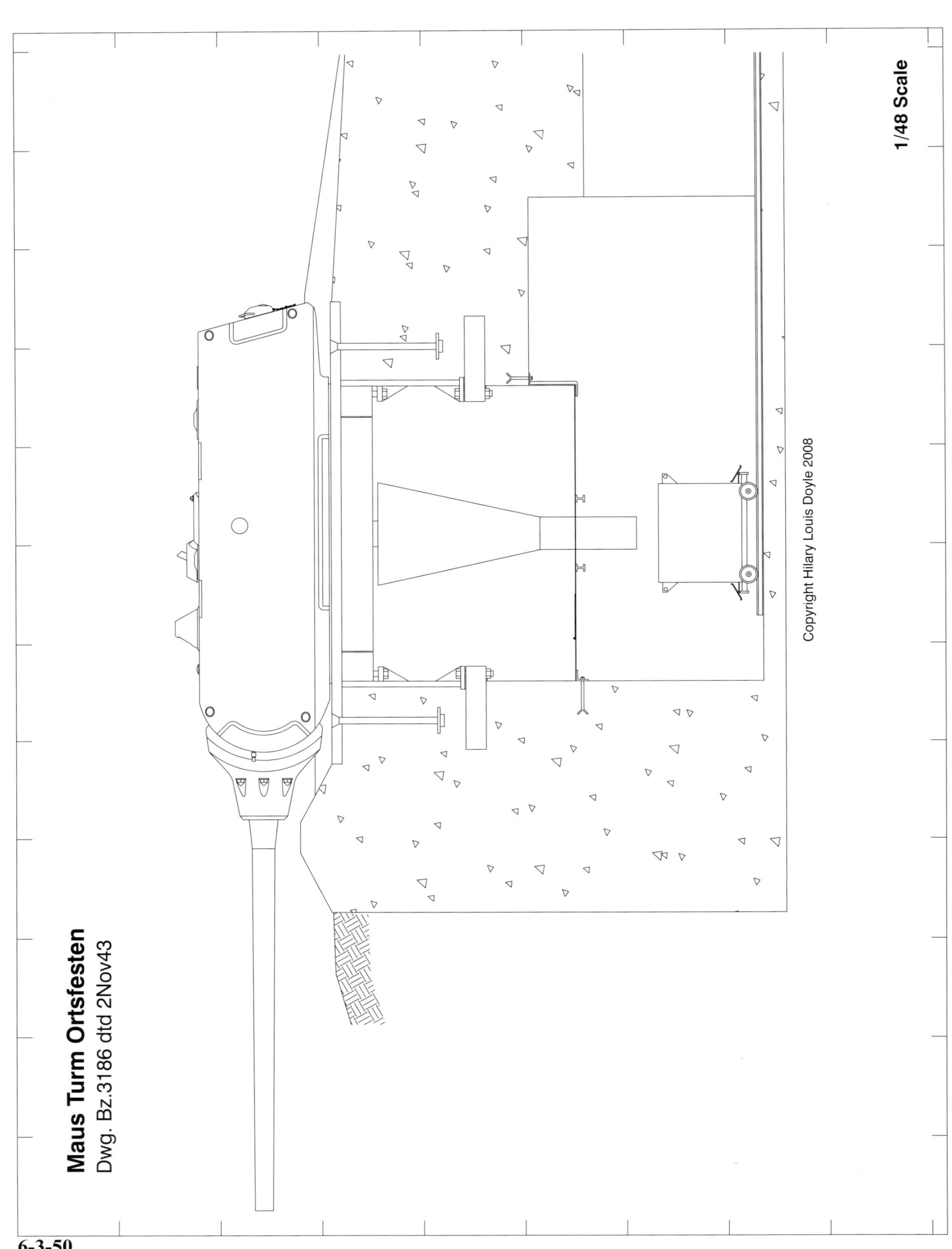

**Maus Turm Ortsfesten**
Dwg. Bz.3186 dtd 2Nov43

# Tiger-Maus, E 100

## KRUPP TIGER-MAUS

The idea of a **150 ton Pz.Kpfw.** as a competitive design to the **Porsche-Maus** was raised during a meeting attended by Generalmajor Fichtner, Oberstlt. Holzhaeuer, and Oberstlt. Krohn (Wa Pruef 6) and Obering. Woelfert (Krupp) on 11 September 1942: *Holzhaeuer asked how far Krupp was involved in the **Porsche 150 ton Pz.Kpfw.** project. Woelfert responded that they were designing the turret. Holzheuer carefully implied that Porsche has difficulties with his **Tiger** because of the numerous new components, and it appears auspicious if others proposed a **150 t Fahrgestell**. Woelfert explained that Krupp was eager to make a conceptual design and asked if information on strong engines and transmissions was available. Holzhaeuer stated that Maybach had promised that the power of their **HL 230 Motor** could be increased to 1000 metric horsepower by using special fuel at increased compression but without a charger. Holz-haeuer wanted to discuss the production of a **150 t Pz.Kpfw.** (other than the Porsche design) in about four weeks with the Panzerkommission and earlier with the Minister fuer Bewaffnung und Munition.*

*Krupp's conceptual design was discussed at a meeting held at the Min.f.Bew.u.Mun. (Odl. Saur) attended by the Panzerkommission (General v. Radelmeier, Prof. Dr.Porsche), In 6 (Oberst Thomale, Oberstlt. Bollbrinker), Pruef 6 (Oberst v. Wilcke, Oberstlt. Crohn, Oberbaurat Kniepkamp) on 17 November 1942. Krupp's proposal is to be quickly submitted. A decision is to be made in 3 or 4 weeks whether the Porsche or Krupp design is accepted for production.*

*Krupp presented a conceptual design with the remark that it was created before the latest requirements had been made. Drawing W 1672 had special **Raupenkaesten** (track boxes) that would have to be dismounted for rail transport. A loading width of 3070 mm would allow opposing rail traffic. Ground pressure was only 0.8 kg/cm2. The newly allowed ground pressure of about 1.1 to 1.2 kg/cm2 was achieved by the design in drawing W 1671. The turret is to be relocated to the rear and the ground pressure lowered from 1.3 to 1.2 kg/cm2.*

*As proposed by Oberstlt. Holzhaeuer, the use of **Henschel-Tiger** components for the mechanical drive train must be determined for shortening the schedule. In this case, the horsepower-to-weight ratio would only be 4.5 for a **155 ton Panzer**. In comparison to the **65 ton Henschel-Tiger** with a maximum speed of 45 km/hr, the maximum speed of a **155 ton Maeuschen** would only be 20 km/hr. There are worries about its climbing ability in difficult terrain.*

*The use of **Henschel-Tiger** drive train components was discussed for the **150 ton Maeschen** (drawing W 1671) on 23 November 1942. In using the same engine as in the **Henschel-Tiger**, (considering the same rolling resistance) calculations result in a maximum speed of 20 km/hr based on weight and only 13 km/hr based on the power needed for the **Henschel-Lenkgetriebe** (steering unit). A new **Lenkgetriebe** is needed that is designed for 800 instead of 360 horsepower so that a Panzer weighing 170 metric tons is capable of 25 km/hr.*

At an internal meeting, Krupp discussed the development of the **Maus** on 25 November 1942. *For this urgent **Maus** development, as many components as possible should be taken from the **R 2** and **R 1** design projects.*
*1. Engines that can be used include the water-cooled Daimler-Benz gasoline **Motor MB 501** rated at 1200/1500 metric horsepower, a water-cooled Daimler-Benz gasoline **Motor MB 503** rated at 1200 horsepower, or the similar Daimler-Benz diesel **Motor MB 507** rated at 800/1000 horsepower at 2200 or 2400 rpm. It is rumored that Porsche is using the Daimler-Benz **Flugmotor DB 603** rated at 1375 horsepower at 2300 rpm for his **Maus**.*

*If there are difficulties meeting deadlines, initially the available Maybach **Motor HL 230** rated at 700 horsepower or Daimler-Benz **MB 507** rated at 800/1000 horsepower can be used.*
*2. The available **Schaltgetriebe** (transmissions) are the Zahnradfabrik **AK 7-200** designed for 800 horsepower with a shifting range of 1:13.4, the Z.F. **Elektromagnetisches Getriebe 12 EV 170** designed for 770 horsepower with a shifting range of 1:15.48 (already installed for testing in a **Henschel Tiger I**), or a Maybach **Olvargetriebe OG 402016** designed for 800 horsepower with a shifting range of 1:16. As newly developing designs there is a stronger Maybach **Olvargetriebe** designed for 1200 horsepower which Wa Pruef 6 especially likes or a Zahnradfabrik **Allklauen-** or **elektromagnetischen Getriebe** which Krupp prefers. Two transmissions are needed by 1 September 1943.*
*3. The **Lenkgetriebe** (steering unit) must be developed by Krupp because there aren't any available with sufficient size. This **Lenkgetriebe** must be designed for a total weight of 170 metric tons, maximum speed of 30 km/hr, 1200 metric horsepower, steering ratio of 1:2, and a shifting range of 1:15.*
*4. The **Seitenantrieb** (final drives) must be newly developed by Gruppe Woelfert.*

*It was decided that the engine remain in the rear. The first conceptual design is to be presented by Dr. Mueller on 28 November 1942.*

As shown in drawing W 1674 dated 30Nov42, the **Maus (170 ton)** had a **Fahrgestell** weighing 122.5 metric tons of which the armor hull weighed 82 tons. It had 32 roadwheels (650 mm diameter) and was powered by a 1200 horsepower **MB 507** engine. The **Turm mit 15 cm Kw.K. L/37** and a **7.5 cm Kw.K. L/24** weighed 47.5 tons with 50 rounds of 15 cm, 100 rounds of 7.5 cm, and 4500 machinegun rounds. It was to be manned by a crew of six.

Krupp's **Maus-Fahrgestell** was discussed at a meeting with Oberbaurat Kniepkamp on 1 December 1942, which resulted in the birth of the **Tiger-Maus** (later renamed **E 100**) design: *Krupp presented drawing W 1674 with the turret in the middle. Removable **Raupenkaesten** (track boxes) make it possible for opposing rail traffic at a loaded width of 3070 mm. Wa Pruef 6 recognized the important advantage of this model when compared to the earlier conceptual design W 1671 with 3700 mm loading width in which opposing rail traffic must be blocked. Wa Pruef 6 held the view that all experience has shown that development of a new steering unit and eventually a transmission would take too long for the especially short deadlines. Also design, production, and testing of removable **Raupenkaesten** would take more time because there isn't any previous experience.*

*Wa Pruef 6 has taken the view that Krupp can produce a **Maus** the fastest, under the presumption that proven drive train components from the **Henschel Tiger II** be used. Krupp was asked for an immediate proposal.*

*In order to use the original Henschel components, the free interior space of 1760 mm was retained. Lengthen the hull and the track contact length as needed for the larger installation diameter of the turret. Wa Pruef 6 proposed the Maybach **Olvar-Getriebe** or Zahnradfabrik **elektromagnetische Getriebe**. Zahnradfabrik informed Krupp that the shifting times were too long for such low power reserves in the vehicle. It was to have a free suspension with rubber-saving roadwheels and separate loading and driving tracks. Ground pressure of about 1.1 kg/cm2 (sunken in). Opposing rail traffic is allowed at a loading width of 3270 mm.*

*By shortening the track contact length and widening the wheel base, it should be possible to use the Henschel **Lenkgetriebe L 801** if the weight of 130 metric tons and a maximum speed of 22 to 25 km/hr is not exceeded. It is necessary to reduce the armor thickness somewhat. Naturally, there is interest in reducing the turret weight.*

*By using the Maybach **Motor HL 230** rated at 700 metric horsepower, the power-to-weight ratio for 130 tons would be about 5.4 horsepower/ton compared to proposal W 1674 at 7 horsepower/ton with 170 tons and 1200 horsepower. The new model has the important advantage that it can be quickly produced and the problems of startup with a newly developed transmission and other components are not expected.*

*Krupp is to present a new design to Wa Pruef 6 at the latest on 8 December.*

On 7 December 1942, Krupp had calculated the maximum speeds for **Maeuschen (130 to)** (drawing W 1677) using the **Henschel-Tiger** drive train. *In using the same engine as the **Henschel-Tiger**, the product of weight and speed must be the same. The maximum speed governed by weight is 22.5 km/hr and maximum speed caused by the **Henschel-Lenkgetriebe** is 21.5 km/hr. With a 1:23.1 gear ratio in the final drive and a maximum speed of 23 km/hr, the **Lenkgetriebe** is overtaxed by 12%.*

By 7 December 1942, Krupp had designed a lighter **130 ton Maus** (drawing W 1677) with a chassis weighing 83.4 tons, hull weighing 52 tons, tracks 1100 mm wide, 32 roadwheels (800 mm dia.) a 700 horsepower **HL 230 Motor**, **Olvar Schaltgetriebe**, **L 801 Lenkgetriebe**, and 1200 liters of fuel. The turret with a 15 cm L/37 and a 7.5 cm L/24 weighed 45.5 tons. A total of 40 rounds were to be stowed for the 15 cm, 75 rounds of 7.5 cm, and 4500 machinegun rounds. It was to be manned by a crew of six.

On 8 December 1942, Obering. Woelfert met with Wa Pruef 6 (Holzhaeuer, Wilcke, Crohn, and Kniepkamp) to present and discuss the **130 ton Maus** (W 1677).

*Krupp's conceptual design W 1677 for a **130 ton Maus** with the turret in the middle and drive train components from the **Henschel-Tiger II** was enthusiastically received. The special advantages in comparison to the **Porsche-Maus** are:*
*1. Steering ratio of 1:1.43 compared to about 1:2.5*
*2. Ground pressure of 1.1 kg/cm² compared to 1:1.27*
*3. Rail travel without blocking the opposing traffic*
*4. At 40 tons lower weight, significantly lower expenditure of raw materials and labor.*
*5. Lower fuel consumption at 130 ton compared to 170.*

*The maximum speed of 23 km/hr and the armor are sufficient. The disadvantages that must be accepted are that a **Verladekette** must be installed for rail transport, the suspension is not protected by armor, and the power-to-weight ratio is only 5.4 horsepower/ton with a 700 horsepower engine. This can later be increased to 7.5 horsepower/ton when the 1000 horsepower Maybach engine with associated transmission is installed, which according to Oberbaurat Kniepkamp should be delivered in September 1943. A stronger final drive and steering unit would need to be developed to match the 1000 to 1100 horsepower.*

*This project would significantly gain favor if the weight was further reduced, especially with a lighter turret. The turret presently weighs 35% of the vehicle in comparison to 17 to 20% for the **Tiger**.*

*Wa Pruef 6 agreed with Krupp's opinion that a turret in the middle is preferable for the immediate design. The comparison drawing W 1679 shows that the achievable slope is significantly better than the **Tiger I** and **II**.*

*The layout with the turret at the rear (drawing W 1681) has the following disadvantages:*
*1. Vehicle center of gravity lies 40 cm further back.*
*2. The **UK-Anlage** (submersion system) with **Teleskoprohr** from the **Tiger** can't be completely adopted.*
*3. **Panther (Tiger II)** engine compartment must be changed.*

4. Heat burden on the crew from the forward engine.
5. Crew separated. The driver is separated from the turret area.

*Kniepkamp will allow Krupp to install a leaf spring or torsion bar suspension, because at low speed the softness of the springs plays only a minor role.*

*There was great interest in quickly getting the **130 ton Maus** into production, which can be immediately mass-produced due to using proven components. Considering Porsche's promises, delivery of the first simplified design **Maus** must be attempted by the Fall of 1943. Because the deadline is based on rapid design, Krupp suggested to Kniepkamp that they get support from Wa Pruef 6 and eventually M.A.N. for details of important suspension components.*

Dr. Mueller and Obering. Woelfert (Krupp) met with Hdl. Saur (Min.f.Bew.u.Mun) on 8 December 1942. *Krupp proposed that a **130 ton Maus** with **Tiger** components be developed in parallel with the **Porsche-Maus**. There are no reservations against the immediate mass production of this model. Saur declared that he was in agreement but would still have to obtain permission from Minister Speer.*

On 15 December 1942, Krupp was informed that work on the weaker **Maus** design with **Tiger** components should cease, because Oberstlt. Holz-haeuer after consulting with Chef H. Rust has decided that only the **Porsche-Maus** with **Krupp-Turm** be produced.

On 17 December 1942, Krupp met with Wa Pruef to discuss the **130 ton Maus** with Tiger components. Oberstlt. *Holzhaeuer related that at a meeting in the Fuehrerhauptquartier on 2 December 1942, Prof. Porsche was authorized to build a **Maus** in accordance with previously presented conceptual designs because he had promised delivery in the Summer of 1943. No decision has been made on the Krupp design for removable **Raupenkaesten** which was presented at the same time. There are doubts whether the new transmission and steering unit can be obtained in time. The Krupp design for a **130 ton Maus** with proven **Tiger** components wasn't presented at this meeting.*

*Wa Pruef 6 has the opinion that production of a **130 ton Maus** with **Tiger** components is extremely useful, but apparently parallel development of two different **Maus** chassis is unfortunately being avoided because of the **Tiger** experience. In regard to the low power of the Maybach HL 230 Motor, Oberbaurat Kniepkamp stated that a super-charged Maybach **Versuchs-Motor** with at least 1000 horsepower is foreseen in September 1943. In addition, a **hydro-mechanisches Schalt- und Lenkgetriebe** is being developed by a team from Zahnradfabrik, Maybach, A.E.G. and Voith*

*Krupp intends to meet with Minister Speer or Hdl. Saur to obtain a development contract for the **130 ton Tiger-Maus**.*

On 31 December 1942, Oberstlt. Holzhaeuer informed Krupp that after consulting with Hdl. Saur, production of the **Krupp Tiger-Maus** was to be proposed to Hitler. He would also travel to the Fuehrerhauptquartier and asked that Dr. Mueller bring along the necessary supporting documentation.

As recorded in Hitler's conference with Speer on 3 to 5 January 1943 as Item 9: *After thoroughly weighing and comparing the advantages and disadvantages of the proposed "Maeuschen" from Krupp and Porsche, Hitler has decided that the Porsche design will be accepted for production.*

## **ADLER E 100**

At a meeting on 18 March 1944, Obering. Woelfert (Krupp) learned about the **Einheitsfahrzeug E 100**. Min.Rat. Kniepkamp (Wa Pruef 6) was informed that the drawings were to be picked up by Herr Halberkamp from Dir. Jenschke (Adler, Frankfurt) and that inspection of the wooden model would follow in about 1 week.

On 17 May 1944, Obering. Rabe (Porsche) reported seeing a turret drawing which had been adopted for the **E 100** and weighed only 35 metric tons. It had a sloped front, and the 7.5 cm gun was mounted above the 12.8 cm gun. The only difference noted between this **E 100 Turm** and the **Maus II Turm** (drawing Bz 3269) was thinner armor plates (200 mm front, 80 mm sides, 150 mm rear, and 40 mm deck).

Ober Ing. Woelfert reported on a meeting held in Kummersdorf on 30/31 May 1944 about the **15 cm auf E 100**: *During a meeting with Min.Rat. Kniepkamp about the further development of the **E 100**, Kniepkamp admitted that the drawing of our **Tiger-Maus** from November 1942 was the basis for the **E 100** and that the only change was the springs for the suspension. After the decision at the end of 1942 in favor of the **Porsche-Maus**, in the Spring of 1943 Kniepkamp had resurrected our project and within the framework of his **Entwicklungsreihe Versuchs-Panzerkampfwagen** had obtained permission from the Panzerkommission through Oberst Holzhaeuer to build one **E 100 Versuchs-Fahrgestell**. When asked why he had given Krupp's design to another firm (Adler) which didn't have any previous experience in designing either turrets or chassis for Panzers, Kniepkamp replied that in his opinion, Krupp was overburdened with other work.*

*I spoke with Oberst Holzhaeuer the next day about our work together with Adler on the **E 100 Fahrgestell**. I explained to him that other than the suspension springs, the current **E 100** was completely based on our **Tiger-Maus** design already presented in November 1942. I had the impression that Oberst Holzhaeuer had been misinformed by Kniepkamp because he believed that the **E 100** was basically different from the **Tiger-Maus**. However I was able to convince Oberst Holzhauer that the opposite* was true.

I also spoke with Oberst Crohn about my discussions with Holzhaeuer and Kniepkamp. Crohn didn't want anything to do with Kniepkamp and his **Entwicklungfahrzeugen**. But he did agree to send our drawings to Oberst Holzhaeuer with a note to prove that Krupp was the creator of the **E 100** design.

As reported postwar in the Development of New Series German Tanks up to end of March 1945 by Major R.E. Kaufman dated 28 August 1945 - "The engineering staff of Adler, under the direction of Dir. Jenschke, had been loaned to the Heeres Waffenamt to design the **E-100** tank. The engineering staff of Adler, working at Friedberg, started the design on 30 June 1943. The **E-100** design was finished and the parts were assembled at Paderborn for construction of a pilot."

On 15 January 1945, a progress report on the **E 100 Fahrgestell** being assembled was sent to Wa Pruef 6 and Adler in Frankfurt from Haustenbeck near Paderborn.

*64 different photographs have been included to provide a general overview of the previously completed work done on the E 100 Fahrgestell.*

*Every photograph in this report has an explanatory caption. Because of the small hall in which this Fahrgestell was built, photographs couldn't be taken with a larger view. I hope that these photographs are sufficient to provide a picture of how far the work has progressed.*

*As a result of many difficulties caused by the war situation, it wasn't always possible to receive parts when needed, resulting in assembly of the Fahrgestell not being further along. In addition, it should be noted that only three employees from Adler, who are kept fully busy, are continuously here to work on assembly.*

*Coil springs for the suspension were sent by rail to the wrong location and still haven't arrived. The Adler employees informed me that the suspension can be completed after these coil springs arrive. The Transportkette (rail transport track) is stored here but the Gefechtskette (combat track) still hasn't arrived.*

*Assembly of those parts located (in the engine compartment) between the firewall and the rear of the hull have been completed, except for the fuel lines which haven't been delivered.*

*Parts needed for the Kampfraum (fighting compartment) and components mounted in it have arrived and are currently being installed. The drive from the Triebrad (drive sprocket wheel), through the Seitenvorgelege (final drive), Bremsen (brakes), Lenkgetriebe (steering unit), Schaltgetriebe (transmission), and Kardanwelle (drive shaft) up to the engine will then be completed.*

*The Abdeckplatte (cover plate) over the transmission and steering unit hasn't arrived yet. The electrical system can also be completed after this cover plate arrives onto which the instrument panel is fastened.*

*After the fuel lines and the electrical system are completed, the drive train can be operated. Henschel, who is rumored to be responsible for delivery of the Abdeckplatte, will be notified by me. A list will be made of other still missing small parts so that the responsible specialists at Adler can be put to work for their rapid delivery.*

*Information about the Turm (turret) and its delivery or the similarly shaped Fahrgewichten (test weight) is requested. This is needed so that transport can be arranged for the apparently very high test weight from the Versuchsgelaende (proving ground) to the assembly hall. When the dimensions and weight are known, an appropriate method can be arranged.*

The **Zusammenstellung "E 100"** overview drawing **021A38300** (redrawn for the Allies after the war by Adler using partially burnt drawings) still has the same turret that was drawn by Krupp in December 1942 for their **Tiger-Maus** upon which the **E-100 Versuchsfahrgestell** was based. This drawing has preserved the features of a turret as designed by Krupp in December 1942 for both the **Porsche-Maus** and their **Tiger-Maus** with a commander's cupola, a 15 cm L/37 gun with muzzle brake, 7.5 cm L/24 gun, vision ports on the sides, and a crew hatch on the rear with a pistol port (before all of these items were changed for the **Maus** turret as requested by Wa Pruef 6 starting in January 1943).

As reported on 17 May 1944, Krupp had designed a new turret for the **E 100** which weighed only 35 metric tons. This **E 100 Turm** had a sloped front and the 7.5 cm gun was mounted above the 12.8 cm gun. The only difference noted between this **E 100 Turm** and the **Maus II Turm** (drawing Bz 3269) was thinner armor plates (200 mm front, 80 mm sides, 150 mm rear, and 40 mm deck) for the **E 100 Turm**.

Still only partially assembled at the end of the war, the single **E 100 Versuchsfahrgestell** had a drive train consisting of a **Maybach HL 230 P30** rated at 700 metric horsepower at 3000 rpm, a **Maybach OG 40 12 16 B Schaltgetriebe** (transmission), and a **Henschel L 801 Zweiradien Lenkgetriebe** (two-radius steering unit). With the new turret the **E 100** weighed a total of about 130 metric tons. Armor plates were 200 mm thick on the upper hull front at 60°, 150 mm lower front hull at 52°, 120 mm hull sides at 0°, 150 mm rear plate at 30°, 40 mm roof, and 80 mm belly forward, 40 mm aft.

The **Project B Antrieb** (drive train) listed on drawing **021A38300** with a Maybach 1200 horsepower engine and an **8-speed "Mekydro"** mechanical/hydraulic combination transmission and steering unit was being designed for a rear drive and capable of a maximum speed of 40 km/hr. This necessitated moving the engine compartment forward and would have resulted in an entirely different hull shape for a future stage in the **E 100** design evolution.

Above and Below: The following photographs (pages 6-3-55 to 6-3-66) were attached to the progress report dated 15 January 1945 on the status of E 100 Fahrgestell (chassis) assembly at Haustenbeck near Paderborn, Germany. (NA)

Above and Below: The drive sprocket wheels on the right and left side with details of the final drive housing, the hull side extensions cut out for towing eyes, and the mounting eye for the track guard. (NA)

Above: The front deck with the cutout for the missing Abdeckplatte (cover plate with the driver's and radio operator's hatches). Inside, two shock absorbers are located on each side with the covered transmission and main drive shaft cover in the middle. Below: Curved segments were cut out of the inside surface of the hull sides to gain clearance for the turret platform. (NA)

This Page and Upper Right:
The driver's seat was adjustable in height to allow driving with the driver's head protruding above the hatch. In addition to the steering wheel, there were two laterals for the brakes as well as parking brake levers. The preselection gear lever for the 8-speed semi-automatic transmission was on the driver's right. (NA)

Right: Two batteries were located in the hull on either side behind the fuel tank. An automatic fire extinguisher, cold-start fuel injector, main electrical power switch, and voltage regulator are mounted on the fire wall. (NA)

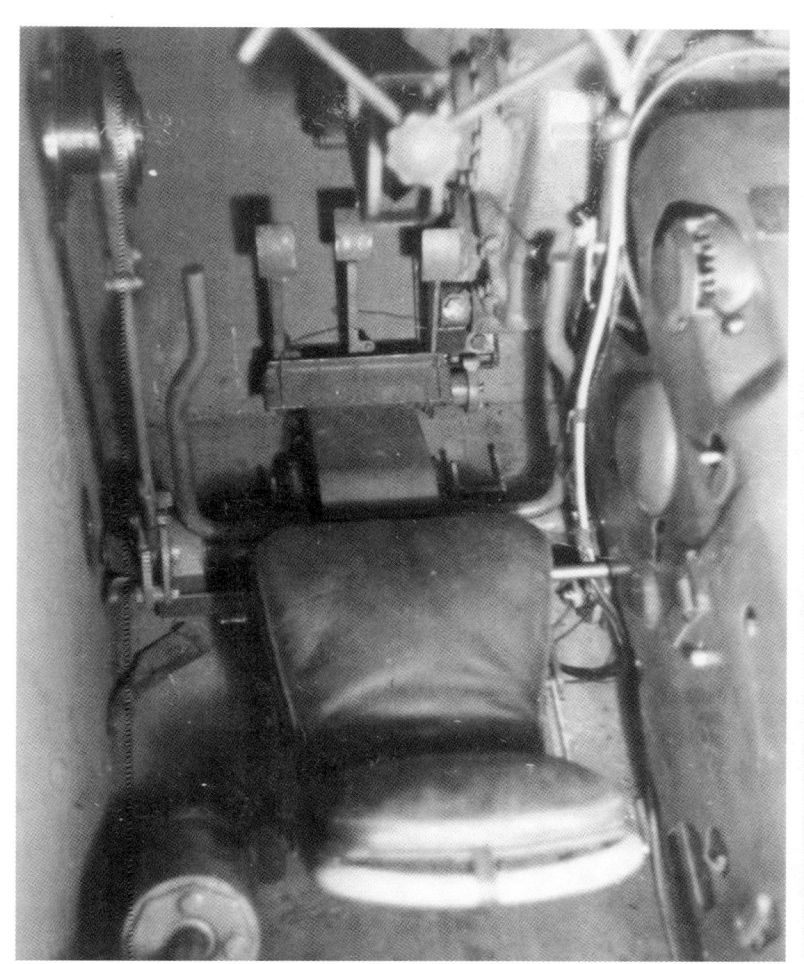

6-3-60

This and Opposite Page: Details of rear deck components, including the engine access hatch, air intake louvers with screens, cooling fan (without the cast louver), (a) combustion air intake (minus the screen), (b) fuel filler cap, and (c) radiator filler cap.

Above: The fuel filler chamber was on the left and the cooling water filler chamber on the right. One of the magnetoes has been removed from the rear of the Maybach HL 230 P30 engine. (NA)
Below: The twin air cleaners mounted on top of the Maybach HL 230 P30 engine. (NA)

**Above and Below:** 900 mm diameter roadwheels with rubber-saving steel tires were mounted on swing arms in pairs. Coil springs for the suspension had been sent by rail to the wrong location and still hadn't arrived when these photographs were taken.

Left and Upper Right:
The idler wheels were mounted on crank arms with the track adjusting mechanism mounted inside the hull. The armor guards have not been mounted on the exhaust tail pipes. A cylindrical Abstandsruecktlicht (convoy tail light) was mounted on the right side of the tail plate. (NA)

Below: This photograph of the 1000 mm wide Gefechtskette (combat track) was taken after the war. Only the Transportkette (rail transport track) was available in January 1945. (TTM)

Below: The two outer holes (with caps missing) in the belly plate were for access to the track adjusting mechanisms.  There was a remotely operated drain valve at the left rear, an access port to the cooling system to the front left, an access hole to the starter on the right rear, and a fuel drain on the right front.

**Above and Below:** All six sections of the Kettenschutz (track guard) had been delivered and were stored outside when the assembly status report was written in January 1945. (NA)

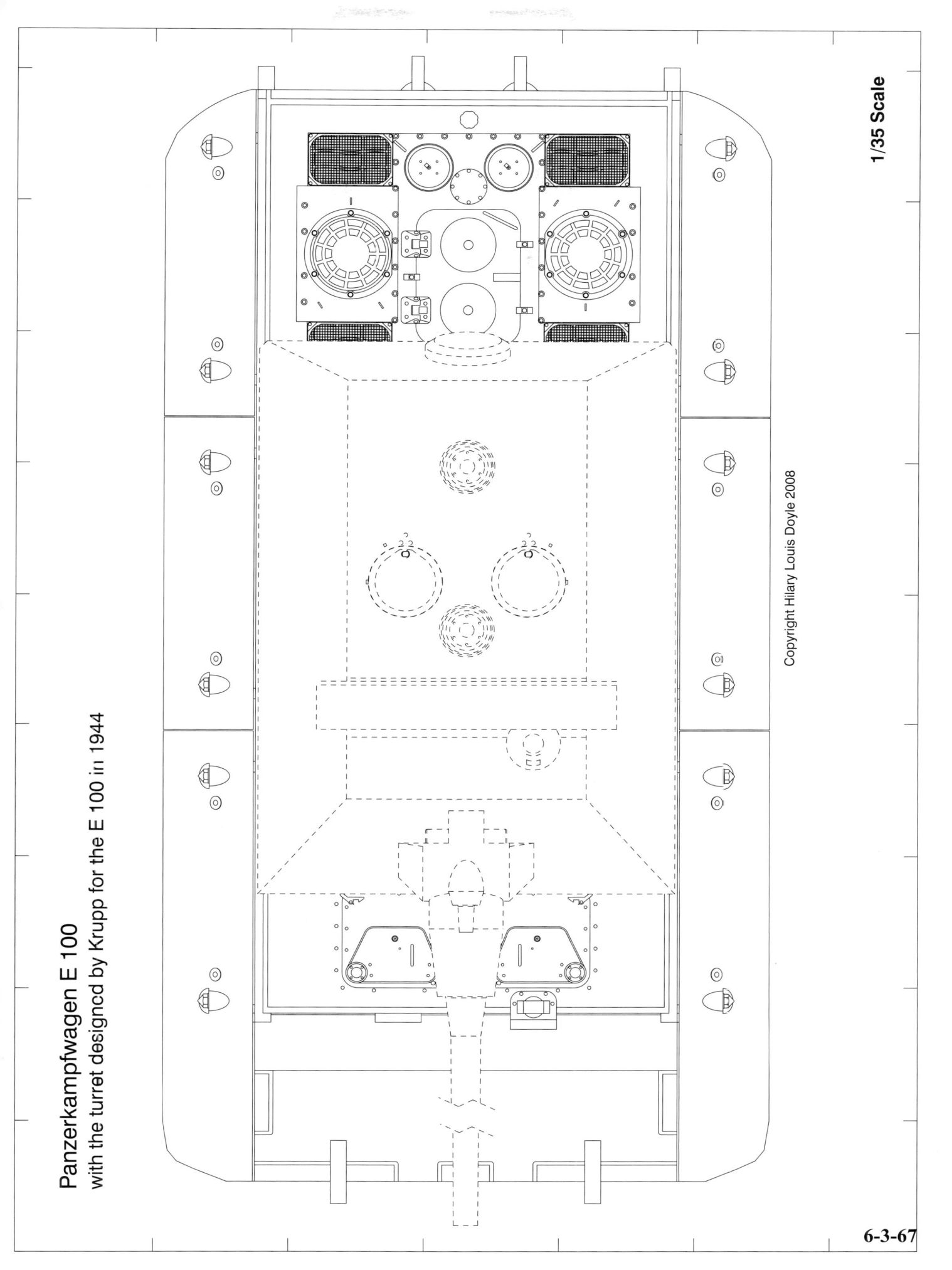

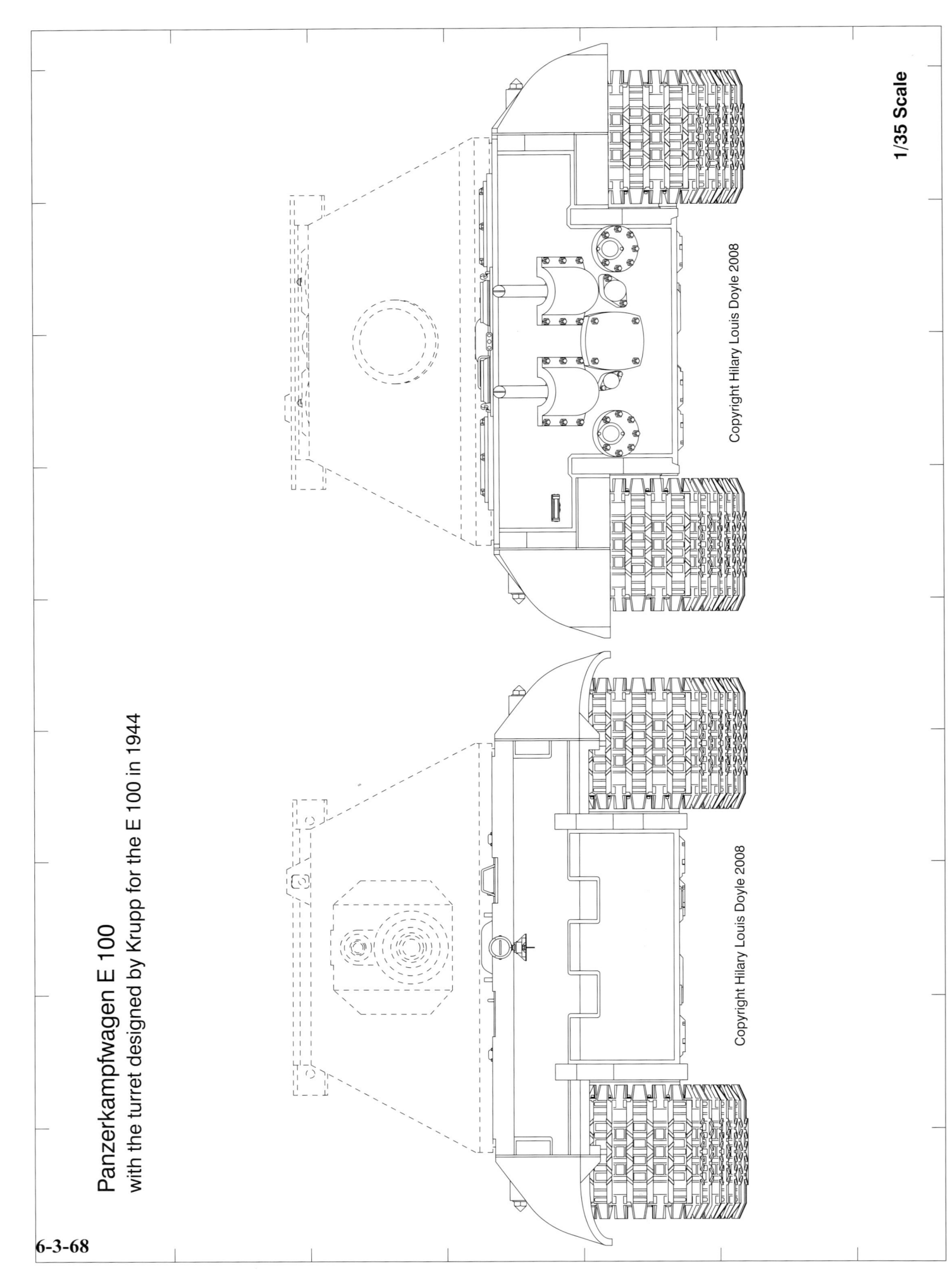

## Panzerkampfwagen E 100
with the turret designed by Krupp for the E 100 in 1944

1/35 Scale

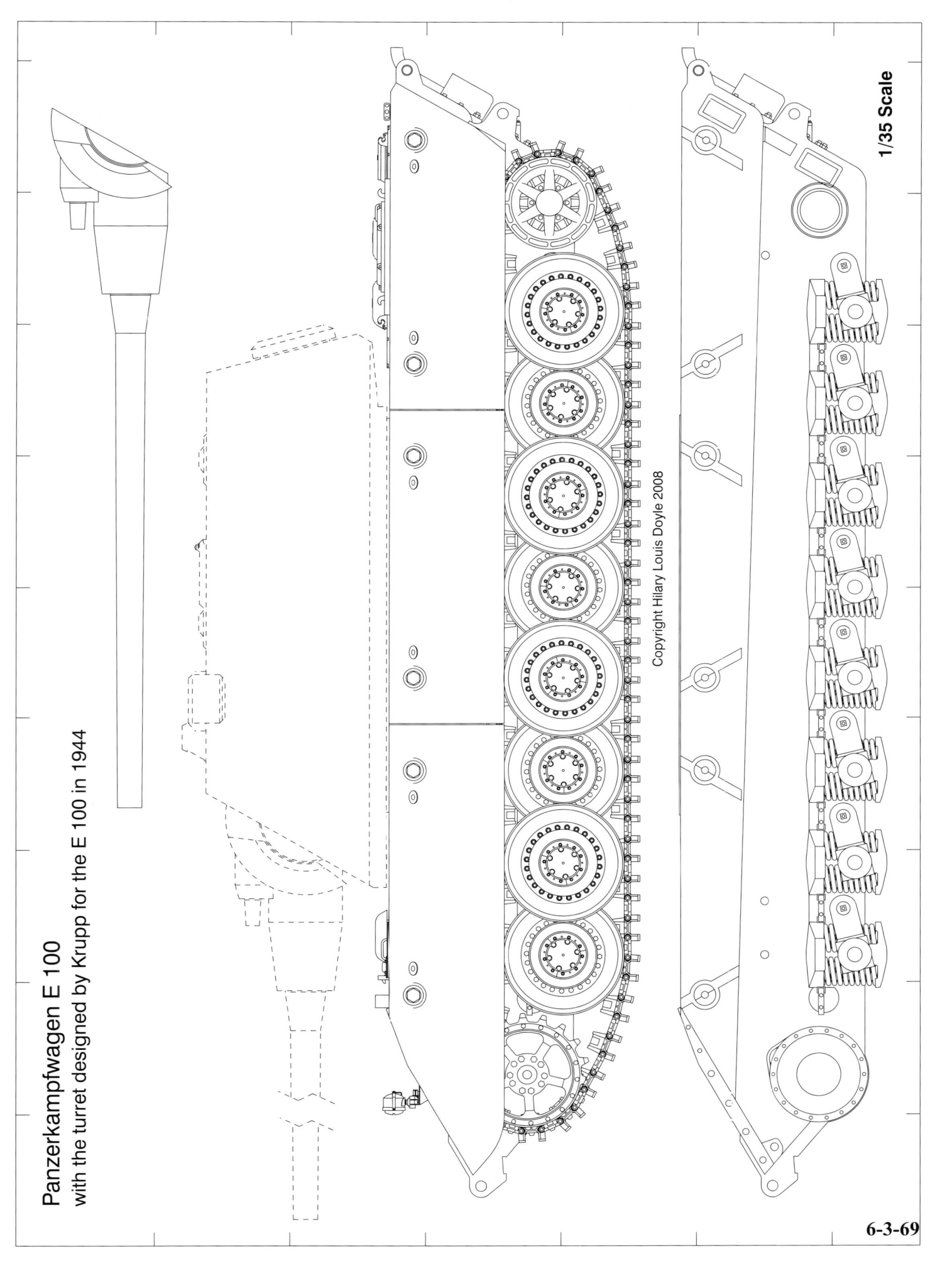

**This Page and Right: The partially assembled E 100 Fahrgestell was moved outside after being captured by the Allies at the end of the War. The drive sprocket rings were still missing. (TTM)**

6-3-71

# Panzerkampfwagen "Tiger-Maus"
# E 100

**Weapons Data:**
In Turret: 1 - 12.8 cm Kw.K. (L/55)
 1 - 7.5 cm Kw.K. (L/24)

Elevation: -7, + 20°
Traverse: 360° (hydraulic and hand)
Gun Sight: T.W.Z.F.1 (3x, 10°)
 graduated to 4000 m for Pzgr.

**Ammunition:** ??

**Crew:**
Pz.-Fuehrer (commander)
Richtschuetze (gunner)
2 Ladeschuetzen (loaders)
Fahrer (driver)
Funker (radio operator)

**Communication:** Fu 5 and intercom

**Measurements:**
Length, overall: 11.073 m
Length, w/o gun: 8.733 m
Width, overall: 4.480 m
Height, overall: 3.375 m
Firing Height: 2.450 m
Wheel Base: 3.075 m
Track Contact: 4.900 m
Combat Loaded: 123.5 metric ton
Fuel Capacity: 2050 liters

**Automotive Capabilities:**
Maximum Speed: 23 km/hr
Avg. Road Speed: ?? km/hr
 Cross Country: ?? km/hr
Range on Road: 160 km
 Cross Country: 100 km
Grade: 30°
Trench Crossing: 2.9 m
Step: 85 cm
Fording Depth: 165 cm
Ground Clearance: 50 cm
Ground Pressure: 1.26 kg/cm$^2$
Power Ratio: 4.8 HP/ton
Steering Ratio: 1.6

**Automotive Components:**
Motor: Maybach HL 230 P30
 V-12, water-cooled
 23 liter gasoline
 600 HP @ 2500 rpm
Transmission: 8 speed OG 40 12 16B
Steering: L 801 double radius
Drive: Front sprocket
Roadwheels: 8 x 2 per side
Tires: 900 mm dia. Steel
Suspension: Coil springs
Track: 1000 mm wide dry pin
 150 mm pitch
Links per side: 53 & 53

## Armor Specifications for the E 100

Armor thickness in mm/angle from vertical

Tolerances for plate thicknesses -0 to +5%

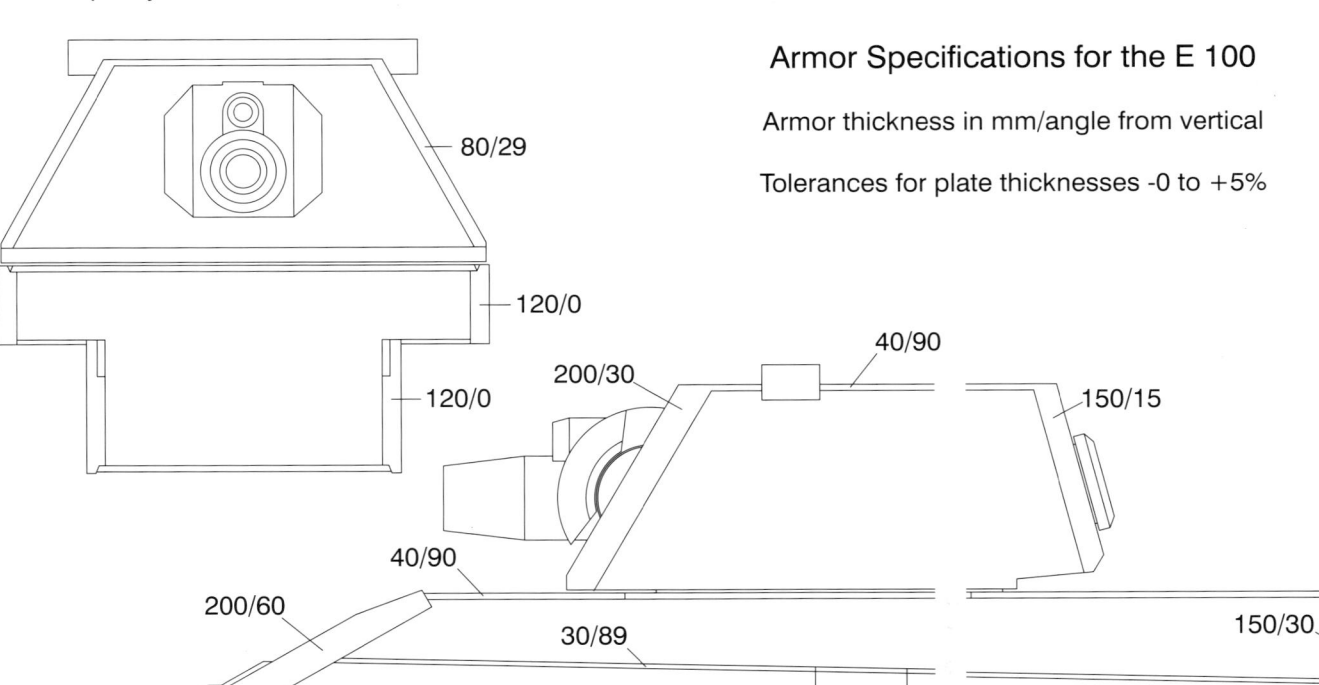

1/48 Scale    Copyright Panzer Tracts 2008